BIBLIOGRAPHY OF EXTRACTIVE INDUSTRIES ACCOUNTING 1960-1982

BIBLIOGRAPHY OF EXTRACTIVE INDUSTRIES ACCOUNTING 1960-1982

Charles E. Boynton, IV
Alan D. Campbell
Editors

Extractive Industries Accounting Research Institute
North Texas State University
P.O. Box 13677
Denton, Texas 76203

Edited by Charles E. Boynton, IV
and Alan D. Campbell

International Standard Book Number: 0-940966-05-0
Library of Congress Catalog Card Number: 84-80466

Extractive Industries Accounting Research Institute
North Texas State University

Printed in the United States of America

TABLE OF CONTENTS

PREFACE

Extractive industries' accounting, particularly oil and gas accounting, has been subject to intensive scrutiny and debate since the early 1960's. In 1964, the Accounting Principles Board (APB) of the American Institute of Certified Public Accountants (AICPA) commissioned research that was published as Accounting Research Study (ARS) No. 11, *Financial Reporting in the Extractive Industries* (Field, 1969). The APB then formed the Committee on Extractive Industries to review ARS No. 11 and recommend a response, including a proposal for an APB Opinion, if it found one was needed. In 1970, the Committee proposed rules for oil and gas accounting. Before action could be taken by the APB, it was replaced by the Financial Accounting Standards Board (FASB) as the authoritative, private sector rule-making body for financial accounting in the U.S. The APB Committee marked the end of its tenure with a final report summarizing its findings and recommendations, *Accounting and Reporting Practices in the Oil and Gas Industry* (AICPA, APB Committee on Extractive Industries, 1973). In the following decade, the FASB issued Financial Accounting Standards No. 9 (1975), No. 19 (1977), No. 25 (1979), No. 39 (1980), No. 69 (1982), and No. 71 (1982) and FASB Interpretations No. 33 (1980) and No. 36 (1981).

The work of the APB Committee and the FASB has involved extensive interaction with the Securities and Exchange Commission (SEC), with industry, and with the public, as this *Bibliography* testifies. The Extractive Industries Accounting Research Institute (EIARI) is publishing this *Bibliography* to document the two-decade period of examination. The editors have chosen to divide the literature into two sub-periods, 1960-1972 and 1973-1982, for three reasons. First, the 1972 and 1973 dates reflect the shift from the APB to the FASB. Second, the dates mark an increase in the public debate over full costing versus successful efforts accounting in oil and gas exploration. Third, the dates approximate the shift to post-OPEC petroleum prices. Those price increases led to a profound increase in government regulation of the petroleum industry and to wide interest in the measurement of petroleum industry profitability.

Although the 1960-1982 period also saw changes in the taxation of extractive industries, this *Bibliography* is limited to the literature addressing the financial accounting, managerial accounting, and auditing problems of the industries.

This *Bibliography* was compiled from a number of sources including the *Accountant's Index*, the *Wall Street Journal Index*, computer data bases, and various industry and industry-related publications. It should be noted that in general the original documents have not been examined or verified.

The EIARI plans to issue a 1983-84 edition of the *Bibliography* in 1985. The editors would appreciate learning of errors and omissions in the 1960-1982 edition. Corrections and additions to this edition will be included in the 1983-84 edition.

Charles E. Boynton, IV
Alan D. Campbell

Denton, Texas
February 1984

AVAILABILITY OF MATERIALS

Members of the AICPA have access to many of the documents through the AICPA Library. Members may write American Institute of CPAs, 1211 Avenue of the Americas, New York, New York 10036-8775 or call the Library toll free (800) 223-4155 outside New York state or at (800) 522-5434 within New York state. Nonmembers may not borrow from the Library without written authorization from a member who guarantees payment for postage, loss, or damage.

ACKNOWLEDGEMENTS

The editors wish to thank Southern Methodist University for allowing access to their Law Library. We also wish to thank John Jolly, executive director of the Council of Petroleum Accountants Societies (COPAS) for providing information on COPAS publications.

The editors appreciate the comments received on the draft of this publication from the following persons: Omer Adams of Atlantic Richfield Company; William E. Boswell of Standard of Ohio; Claude W. Dodgen, Jr., of Texas Pacific Oil Company; R. C. Drummond of Mobil Oil Corporation; Marcy D. Echols of ARCO Exploration Company; John C. Linehan of Getty Oil Company; J. E. Niemeyer of Enserch Exploration; Eugene M. Parrish of The Louisiana Land and Exploration Company; T. R. Pickrell of Conoco; and E. L. Ramirez of Moore McCormack Energy.

The editors acknowledge the work of Jackie Burrow and Patsy Lee, graduate students in accounting at North Texas State University, who took part in the preliminary compilation phases. The hard work of Sue Zant and Regina Smith made it possible to convert this bibliography from a collection of index cards to a finished work ready for publication.

ORGANIZATION OF THE BIBLIOGRAPHY

The *Bibliography* is divided into two main time periods, 1960 to 1972 and 1973 to 1982. Within each time period material has been classified as oil and gas or mining.

The oil and gas divisions have been subdivided into categories according to the type of publication. Three publication categories (proceedings, news releases and short comments, and articles) have been still further divided into subject categories. News releases and short comments appear in chronological order with each subject category. All other citations are in alphabetical order by author within each subject category if subject categories are used and otherwise within each publication category.

Because there are fewer citations in the mining divisions, they have not been subdivided by type of publication, but are subdivided by subject category. Citations other than news releases and short comments are in alphabetical order by author within each subject category. New releases and short comments, if they occur, appear in chronological order at the end of each subject category.

1960-1972

Oil & Gas

OIL AND GAS
PRONOUNCEMENTS AND REPORTS OF NONGOVERNMENTAL ACCOUNTING AND AUDITING STANDARDS BODIES
1960-1972

Field, Robert E. Financial Reporting in the Extractive Industries. Accounting Research Study No. 11. New York: American Institute of Certified Public Accountants, Accounting Research Division, 1969.

Lourens, Roy, and Henderson, Susan. Financial Reporting in the Extractive Industries: An Australian Survey. Melbourne: Australian Society of Accountants, 1972.

OIL AND GAS GOVERNMENT DOCUMENTS 1960-1972

New York. Public Service Commission. Department of Public Service Uniform System of Accounts for Gas Corporations, Classes A and B. Albany, 1964.

New York. Public Service Commission. Department of Public Service Uniform System of Accounts for Gas Corporations, Classes C and D. Albany, 1965.

New York. Public Service Commission. Department of Public Service Uniform System of Accounts for Gas Corporations, Classes C and D. Albany, 1965.

New York. Public Service Commission. Uniform System of Accounts for Gas Corporations, Class A and B; as Contained in the Official Compilation of Codes, Rules and Regulations of the State of New York, Title 16, Sections 310.1 Through 317.7 and Accounts. Following, Together with Appendixes 6B and 6C. Albany, 1963.

U.S. Federal Power Commission. Depreciation Practices of Natural Gas Companies, 1969, Classes A and B. Washington: Government Printing Office, 1971.

U.S. Federal Power Commission. Natural Gas Act, April 1, 1965. Washington: Government Printing Office, 1965.

U.S. Federal Power Commission. Regulations to Govern The Preservation of Records of Natural Gas Companies, Effective December 12, 1962. Washington: Government Printing Office, 1963.

U.S. Federal Power Commission. Revisions in Uniform Systems of Accounts for Natural Gas Companies (Classes A, B, C, and D) and Annual Report Form No. 2. Washington: Government Printing Office, 1971.

U.S. Federal Power Commission. Rule of Practice and Procedure (18 Code of Federal Regulations, Chapter 1, Parts 1, 2, and 3), in Effect on January 1, 1966. Washington: Government Printing Office, 1965.

U.S. Federal Power Commission. Statistics of Natural Gas Companies 1961; Compiled From Annual Reports Submitted by Natural Gas Companies Pursuant to the Requirements of the Natural Gas Act--Classes A and B Companies. Washington: Government Printing Office, 1962.

U.S. Federal Power Commission. Statistics for Interstate Natural Gas Pipeline Companies 1963; Compiled from Annual Reports Submitted by Interstate Nautral Gas Pipeline Companies Pursuant to the Requirements of the Natural Gas Act--Classes A and B Companies. Washington: Government Printing Office, 1964.

U.S. Federal Power Commission. Statistics for Interstate Natural Gas Pipeline Companies 1964; Compiled from Annual Reports Submitted by Interstate Natural Gas Pipeline Companies Pursuant to the Requirements of the Natural Gas Act--Classes A and B Companies. Washington: Government Printing Office, 1965.

U.S. Federal Power Commission. Statistics for Interstate Natural Gas Pipeline Companies 1965; Compiled from Annual Reports Submitted by Interstate Natural Gas Pipeline Companies Pursuant to the Requirements of the Natural Gas Act--Classes A and B Companies. Washington: Government Printing Office, 1966.

U.S. Federal Power Commission. Statistics for Interstate Natural Gas Pipeline Companies 1966; Compiled from Annual Reports Submitted by Interstate Natural Gas Pipeline Companies Pursuant to the Requirements of the Natural Gas Act--Classes A and B Companies. Washington: Government Printing Office, 1967.

U.S. Federal Power Commission. Statistics for Interstate Natural Gas Pipeline Companies 1967; Compiled from Annual Reports Submitted by Interstate Natural Gas Pipeline Companies Pursuant to the Requirements of the Natural Gas Act--Classes A and B Companies. Washington: Government Printing Office, 1968.

U.S. Federal Power Commission. "Uniform System of Accounts for Class C and Class D Natural Gas Companies." Federal Register, 25 (August 17, 1960) Part II, 7917-7970.

U.S. Federal Power Commission. Uniform System of Accounts Prescribed for Natural Gas Companies (Class A and Class B) Subject to the Provisions of the Natural Gas Act, Effective January 1, 1961. Washington: Government Printing Office, 1960.

U.S. Federal Power Commission. Uniform System of Accounts Prescribed for Natural Gas Companies (Class C) Subject to the Provisions of the Natural Gas Act, Effective January 1, 1961. Washington: Government Printing Office, 1961.

U.S. Federal Power Commission. Uniform System of Accounts Prescribed for Natural Gas Companies (Class D) Subject to the Provisions of the Natural Gas Act, Effective January 1, 1961. Washington: Government Printing Office, 1961.

U.S. Federal Power Commission. Uniform System of Accounts Prescribed for Natural Gas Companies (Class A and B) Subject to the Provisions of the Natural Gas Act, in Effect on February 1, 1964. Washington: Government Printing Office, 1964.

U.S. Federal Power Commission. Uniform System of Accounts Prescribed for Natural Gas Companies (Class A and B, C, and D) Subject to the Provisions of the Natural Gas Act, in Effect on September 1, 1968. Washington: Government Printing Office, 1968.

United States. Federal Utility Regulation Annotated: Supplement D to Volume 2, Federal Power Act and Natural Gas Act. Washington: Public Utilities Reports, 1969.

U.S. Securities and Exchange Commission. Regulation of Oil and Gas Programs. Washington: Government Printing Office, 1972.

OIL AND GAS
COPAS BULLETINS
1960-1972

Council of Petroleum Accountants Societies. "Accounting for Farmouts/Farmins, Net Profits Interests, and Carried Interests." COPAS Bulletin No. 9. Tulsa: Kraftbilt, 1969.

Council of Petroleum Accountants Societies. "Accounting for Unitizations." COPAS Bulletin No. 11. Tulsa: Kraftbilt, 1971.

Council of Petroleum Accountants Societies. "Accounting Procedure, Joint Operations." COPAS Bulletin No. 5. Tulsa: Kraftbilt, 1966.

Council of Petroleum Accountants Societies. "Accounting Procedure, Joint Operations." COPAS Bulletin No. 8. Tulsa: Kraftbilt, 1969.

Council of Petroleum Accountants Societies. "Classifications for Use in Summary Form Billing, Producing and Gasoline Plant Operations." COPAS Bulletin No. 1. Tulsa: Kraftbilt, 1963.

Council of Petroleum Accountants Societies. "COPAS Forms." COPAS Bulletin No. 4. Tulsa: Kraftbilt, 1966.

Council of Petroleum Accountants Societies. "Determination of Values for Well Cost Adjustments--Joint Operations." COPAS Bulletin No. 2. Tulsa: Kraftbilt, 1965.

Council of Petroleum Accountants Societies. "Gas Accounting Manual." COPAS Bulletin No. 7. Tulsa: Kraftbilt, 1968.

Council of Petroleum Accountants Societies. "Material Classification Manual." COPAS Bulletin No. 6. Tulsa: Kraftbilt, 1967.

Council of Petroleum Accountants Societies. "Petroleum Industry Accounting Educational Training Guide." COPAS Bulletin No. 10. Tulsa: Kraftbilt, 1971.

Council of Petroleum Accountants Societies. "The Initiation of Joint Account Audits, Protocol and Guides in the Petroleum Industry." COPAS Bulletin No. 3. Tulsa: Kraftbilt, 1965.

OIL AND GAS PROCEEDINGS 1960-1972

GENERAL

Barr, Andrew. "Financial Accounting and Reports to Shareholders From the Viewpoint of the SEC," in Institute on Oil and Gas Accounting, 1st, Southwestern Legal Foundation, 1965 Proceedings. New York: Matthew Bender & Company, 1966, 15-40.

Bass, Jarman. "Generally Accepted Accounting Principles in the Petroleum Industry: What GAAP?" in Accounting Papers, 23rd Conference of Accountants. University of Tulsa, 1969, 81-90.

Dodson, Charles R. "Use of Engineering Reports in the Analysis of Financial Statements for Producing Oil and Gas Companies," in Exploration and Economics of the Petroleum Industry, 3rd, Southwestern Legal Foundation, 1965 Proceedings. Houston: Gulf Publishing Company 1965, 51-67.

Edwards, J. O. "Financial Accounting and Reports to Shareholders From the Viewpoint of Management," in Institute on Oil and Gas Accounting, 1st, Southwestern Legal Foundation, 1965 Proceedings. New York: Matthew Bender & Company, 1966, 41-62.

Field, Robert E. "A Report on the AICPA Survey of Accounting Practices in the Extractive Industries," in Institute on Oil and Gas Accounting, 2nd, Southwestern Legal Foundation, 1966 Proceedings. New York: Matthew Bender & Company, 1967, 149-177.

Field, Robert E. "Financial Reporting in the Extractive Industries--A Synopsis of the Research Study Prepared for the AICPA." in Accounting Papers, 23rd Conference of Accountants. University of Tulsa, 1969, 46-65.

Matthews, A. Bruce. "Accounting Problems of Petroleum and Mining Industries," in Rocky Mountain Mineral Law Institute, 9th, 1963 Proceedings. New York: Matthew Bender & Company, 1964, 205-224.

Norr, David. "Shareholder Reporting in the Oil Industry," in Institute on Oil and Gas Accounting, 1st, Southwestern Legal Foundation, 1965 Proceedings. New York: Matthew Bender & Company, 1966, 1-13.

Norwood, Fred W. "Acquisitions--Pooling Versus Purchases," in Institute on Oil and Gas Accounting, 1st, Southwestern Legal Foundation, 1965 Proceedings. New York: Matthew Bender & Company, 1966, 77-96.

Porter, Stanley P. "Interpretation of the Financial Statement," in Exploration and Economics of the Petroleum Industry, 3rd, Southwestern Legal Foundation, 1965 Proceedings. Houston: Gulf Publishing Company, 1965, 21-50.

Rain, Robert E., Jr. "A Further Look at Division Orders and Problems in Accounting and Payment of Proceeds from Oil and Gas," in Rocky Mountain Mineral Law Institute, 8th, 1962 Proceedings. New York: Matthew Bender & Company, 1963, 69-92.

Reed, W. Rowland. "Problems in Cost Measurement in the Extractive Industries." in Accounting Papers, 26th Conference of Accountants. University of Tulsa, 1972, 36-48.

Richards, L. J. "Top Management Views the Accounting Function," in Institute on Oil and Gas Accounting, 2nd, Southwestern Legal Foundation, 1966 Proceedings. New York: Matthew Bender & Company, 1967, 51-66.

Sandison, Robert W. "Interpreting Financial Statements of a Producing Company." in National Institute for Petroleum Landmen, 6th, Southwestern Legal Foundation, 1964 Proceedings. New York: Matthew Bender & Company, 1965, 183-215.

Spriggs, Dillard P. "Shareholder Reporting of Foreign Oil Operations," in Institute on Oil and Gas Accounting, 2nd, Southwestern Legal Foundation, 1966 Proceedings. New York: Matthew Bender & Company, 1967, 67-88.

Williams, S. D. "Accounting for Exploration Costs," in Institute on Oil and Gas Accounting, 2nd, Southwestern Legal Foundation, 1966 Proceedings. New York: Matthew Bender & Company, 1967, 179-206.

DISCLOSURES AND RESERVE RECOGNITION ACCOUNTING

Keplinger, C. H. "Case Histories of Actual Performance vs. Appraised Prognostication--Petroleum Reservoirs," in Exploration and Economics of the Petroleum Industry, 5th, Southwestern Legal Foundation, 1967 Proceedings. Houston: Gulf Publishing Company, 1967, 281-307.

DRILLING FUNDS, PROGRAMS, JOINT VENTURES AND PARTNERSHIPS

Cook, Clorval A. "The Accounting Procedure of Joint Operating Agreements," in National Institute for Petroleum Landmen, 12th, Southwestern Legal Foundation, 1971 Proceedings. New York: Matthew Bender & Company, 1971, 193-223.

Cunningham, R. M. "Oil and Gas Accounting Procedure: As Viewed by a Joint Interest Operations Director," in Rocky Mountain Mineral Law Institute, 13th, 1967 Proceedings. New York: Matthew Bender & Company, 1967, 397-410.

Ellis, Kirby. "Production of Gas from Joint Interest Properties," in Institute on Oil and Gas Law and Taxation, 21st, Southwestern Legal Foundation, 1970 Proceedings. New York: Matthew Bender & Company, 1971, 47-81.

Farrington, Myrle Holt. "Analysis of the Controversial Provisions of an Operating Agreement," in National Institute for Petroleum Landmen, 11th, Southwestern Legal Foundation, 1970 Proceedings. New York: Matthew Bender & Company, 1970, 77-109.

Fawcett, Dwight W. "Applicability of Federal and State Securities Acts to Sales of Oil and Gas Interests," in Institute on Oil and Gas Law and Taxation, 17th, Southwestern Legal Foundation, 1966 Proceedings. New York: Matthew Bender & Company, 1966, 245-292.

Goodson, J. A. "Accounting for Oil and Gas Joint Ventures," in Institute on Oil and Gas Accounting, 2nd, Southwestern Legal Foundation, 1966 Proceedings. New York: Matthew Bender & Company, 1967, 207-236.

Kenneth, C. M. "Joint Venture Accounting, a la COPAS-1962," in National Institute for Petroleum Landmen, 6th, Southwestern Legal Foundation, 1964 Proceedings. New York: Matthew Bender & Company, 1965, 157-182.

Mosburg, Lewis G., Jr. "Mechanics of Registered and Unregistered Drilling Programs," in Institute on Oil and Gas Law and Taxation, 18th, Southwestern Legal Foundation, 1967 Proceedings. New York: Matthew Bender & Company, 1967, 97-179.

Shrene, Charles E. "A Resume of the Federal Securities Laws as Applied to Oil and Gas Financing," in Institute on Oil and Gas Law and Taxation, 19th, Southwestern Legal Foundation, 1968 Proceedings. New York: Matthew Bender & Company, 1968, 209-237.

Tarner, Jack. "Joint Operating Agreements," in National Institute for Petroleum Landmen, 4th, Southwestern Legal Foundation, 1962 Proceedings. New York: Matthew Bender & Company, 1963, 89-115.

ECONOMICS, FINANCE, AND EVALUATION

Beatty, John R. A. "Selected Problems in Oil and Gas Financing," in Rocky Mountain Mineral Law Institute, 11th, 1965 Proceedings. New York: Matthew Bender & Company, 1966, 79-121.

Cossey, Charles W. "Financing Oil and Gas Exploration--Past Present, and Future," in Institute on Oil and Gas Law and Taxation, 24th, Southwestern Legal Foundation, 1972 Proceedings. Ne.. York: Matthew Bender & Company, 1973, 135-166.

Davis, Warren B. "Enigma of Oil and Gas Finding Costs," in Accounting Papers, 18th Conference of Accountants, University of Tulsa, 1964, 11-34.

Dodson, Charles R. "Use of Engineering Reports in the Analysis of Financial Statements for Producing Oil and Gas Companies," in Exploration and Economics of the Petroleum Industry, 3rd, Southwestern Legal Foundation, 1965 Proceedings. Houston: Gulf Publishing Company 1965, 51-67.

Fawcett, Dwight W. "Applicability of Federal and State Securities Acts to Sales of Oil and Gas Interests," in Institute on Oil and Gas Law and Taxation, 17th, Southwestern Legal Foundation, 1966 Proceedings. New York: Matthew Bender & Company, 1966, 245-292.

Fish, Leonard W. "Models for the Gas Industry." in Papers and Presentations, Regulatory Information Systems Conference. Jefferson City, Mo., 1972, 325-330.

Gardner, William B., Jr. "Valuation of Oil-Producing Properties," in Institute for Tax Assessors, 6th. Austin: University of Texas, Institute of Public Affairs, 1964, 82-95.

Gueymard, A. G. "The Seventies--An Era of Transition for Petroleum & Minerals Financing," in Rocky Mountain Mineral Law Institute, 17th, 1971 Proceedings. New York: Matthew Bender & Company, 1972, 557-578.

Hill, Kenneth E. "Important Factors in Evaluating Petroleum Companies," in Exploration and Economics of the Petroleum Industry, 5th, Southwestern Legal Foundation, 1967 Proceedings. Houston: Gulf Publishing Company, 1967, 253-280.

Hulsey, Bob T. H. "Rate of Return--An Investment Tool," in Exploration and Economics of the Petroleum Industry, 2nd, Southwestern Legal Foundation, 1964 Proceedings. Houston: Gulf Publishing Company, 1964, 17-24.

Jessen, F. W. "Valuation of Gas Producing Properties," in Institute for Tax Assessors, 6th. Austin: University of Texas, Institute of Public Affairs, 1964, 72-81.

Keplinger, C. H. "Case Histories of Actual Performance vs. Appraised Prognostication--Petroleum Reservoirs," in Exploration and Economics of the Petroleum Industry, 5th, Southwestern Legal Foundation, 1967 Proceedings. Houston: Gulf Publishing Company, 1967, 281-307.

Kirkpatrick, B. P., and Rickard, H. E. "Capital Requirements of Producing," on Exploration and Economics of the Petroleum Industry, 3rd, Southwestern Legal Foundation, 1965 Proceedings. Houston: Gulf Publishing Company, 1965, 5-20.

Kleinman, Harold F. "Acquisition and Financing Techniques: Securities, Listing, and Stockholder Considerations," in Institute on Oil and Gas Law and Taxation, 20th, Southwestern Legal Foundation, 1969 Proceedings. New York: Matthew Bender & Company, 1969, 265-281.

McCoy, Charles F. "The Modern Approach to Oil and Gas Financing," in Rocky Mountain Mineral Law Institute, 15th, 1969 Proceedings. New York: Matthew Bender & Company, 1969, 109-138.

Miranda, Albert N., Jr. "Valuation of Oil and Gas Properties," in Tulane Tax Institute, 18th. New Orleans, 1968, 285-301.

Monteith, Edward E., Jr. "Financial Criteria of Oil and Gas Lending," in Institute on Oil and Gas Law and Taxation, 17th, Southwestern Legal Foundation, 1966 Proceedings. New York: Matthew Bender & Company, 1966, 293-301.

Nelson, R. A. "The Rate of Return Method of Measuring Profitability of Investments," in National Institute for Petroleum Landmen, 4th, Southwestern Legal Foundation, 1962 Proceedings. New York: Matthew Bender & Company, 1963, 173-199.

Shrene, Charles E. "A Resume of the Federal Securities Laws as Applied to Oil and Gas Financing," in Institute on Oil and Gas Law and Taxation, 19th, Southwestern Legal Foundation, 1968 Proceedings. New York: Matthew Bender & Company,

Timm, Bert C. "Management Tools for Optimum Allocation of Capital in Petroleum Exploration," in Exploration and Economics of the Petroleum Industry, 6th, Southwestern Legal Foundation, 1968 Proceedings, New York: Matthew Bender & Company, 1968, 135-152.

Tippit, John H. "Financing the Exploration for Oil and Gas," in Rocky Mountain Mineral Law Institute, 9th, 1963 Proceedings. New York: Matthew Bender & Company, 1964, 297-337.

Vance, Harold. "Analytical Appraisals of Oil and Gas Properties," in Institute for Tax Assessors, 3rd. Austin: University of Texas, Institute of Public Affairs, 1961, 147-157.

Wilson, Wallace W., and Redding, John A. "Sources of Capital for Oil and Gas Producers," in National Institute for Petroleum Landmen, 6th, Southwestern Legal Foundation, 1964 Proceedings. New York: Matthew Bender & Company, 1965, 217-234.

PLANNING, CONTROL, AND BUDGETING

Blauvelt, Howard W. "Measurement and Control of Profits for Departments or Cost Centers--Part I," in Institute on Oil and Gas Accounting, 1st, Southwestern Legal Foundation, 1965 Proceedings. New York: Matthew Bender & Company, 1966, 97-116.

Bumgarner, John C. "Measurement and Control of Profits for Departments or Cost Centers--Part II," in Institute on Oil and Gas Accounting, 1st, Southwestern Legal Foundation, 1965 Proceedings. New York: Matthew Bender & Company, 1966, 117-135.

Fish, Leonard W. "Models for the Gas Industry." in Papers and Presentations, Regulatory Information Systems Conference. Jefferson City, Mo., 1972, 325-330.

Franzoni, Robert A., and Burk, Richard O. "Applications of Operations Research to the Petroleum Industry," in Institute on Oil and Gas Accounting, 1st, Southwestern Legal Foundation, 1965 Proceedings. New York: Matthew Bender & Company, 1966, 183-236.

Koons, Robert L. "Measurement and Control of Profits for Departments or Cost Centers--Part III," in Institute on Oil and Gas Accounting, 1st, Southwestern Legal Foundation, 1965 Proceedings. New York: Matthew Bender & Company, 1966, 137-152.

Rain, Robert E., Jr. "A Further Look at Division Orders and Problems in Accounting and Payment of Proceeds from Oil and Gas," in Rocky Mountain Mineral Law Institute, 8th, 1962 Proceedings. New York: Matthew Bender & Company, 1963, 69-92.

Ream, L. M., Jr. "Investigation Preceding Acquisition," in Institute on Oil and Gas Accounting, 1st, Southwestern Legal Foundation, 1965 Proceedings. New York: Matthew Bender & Company, 1966, 63-75.

Reed, W. Rowland. "Problems in Cost Measurement in the Extractive Industries." in Accounting Papers, 26th Conference of Accountants. University of Tulsa, 1972, 36-48.

Timm, Bert C. "Management Tools for Optimum Allocation of Capital in Petroleum Exploration," in Exploration and Economics of the Petroleum Industry, 6th, Southwestern Legal Foundation, 1968 Proceedings, New York: Matthew Bender & Company, 1968, 135-152.

Toll, Daniel R. "Appraising the Results of Decisions," in Accounting Papers, 16th Conference of Accountants, University of Tulsa, 1962, 55-81.

Woollen, R. W. "Budgeting Capital Expenditures for an Integrated Oil Company," in Institute on Oil and Gas Accounting, 1st, Southwestern Legal Foundation, 1965 Proceedings. New York: Matthew Bender & Company, 1966, 153-182.

INTERNATIONAL

Spriggs, Dillard P. "Shareholder Reporting of Foreign Oil Operations," in Institute on Oil and Gas Accounting, 2nd, Southwestern Legal Foundation, 1966 Proceedings. New York: Matthew Bender & Company, 1967, 67-88.

ELECTRONIC DATA PROCESSING

Hodgson, R. N. "Data Processing in the Petroleum Industry," in Accounting Papers, 18th Conference of Accountants, University of Tulsa, 1964, 45-49.

REFINING

Woollen, R. W. "Problems in Petrochemical Accounting," in Accounting Papers, 18th Conference of Accountants, University of Tulsa, 1964, 51-59.

PIPELINING AND TRANSMISSION

Allen, C. M. "Let's Look at Accounting Through a Natural Gas Company Pipeline," in Institute on Oil and Gas Accounting, 2nd, Southwestern Legal Foundation, 1966 Proceedings. New York: Matthew Bender & Company, 1967, 131-148.

Mosburg, Lewis G., Jr. "Use of Gas Producer Costs in Federal Power Commission Rate Proceedings," in Institute on Oil and Gas Accounting, 2nd, Southwestern Legal Foundation, 1966 Proceedings. New York: Matthew Bender & Company, 1967, 89-130.

OIL AND GAS
HEARINGS, LECTURES, SPEECHES
1960-1972

Arthur Andersen & Company. Accounting and Reporting Practices in the Petroleum Industry. Before the Accounting Principles Board. Cases in Public Accounting Practice, Vol. 10. Chicago, 1972.

Arthur Andersen & Company. Accounting for Oil and Gas Exploration and Development Costs. Before the Committee on Extractive Industries of the Accounting Principles Board, November 22-33, 1971. Chicago, 1971.

Bennett, E. D., ed. Petroleum Accounting Lecture Series. Studies in Accounting, No. 3 Austin: Bureau of Business Research, University of Texas at Austin, 1969.

Porter, Stanley P. Full Cost Accounting--the Problem It Poses for Extractive Industries. New York: Arthur Young & Company, 1972.

Stevenson, R. E. "Problems in Accounting Principles for the Petroleum Industry," in D. R. Scott Memorial Lectures, 1968, 81-92.

Tevelow, Rosemarie. Accounting Issues: Natural Gas Industry. Before American Gas Association Financial Forum, May 28, 1971. New York: First Manhattan Company, 1971.

OIL AND GAS
NEWS RELEASES AND SHORT COMMENTS
1960-1972

GENERAL

"How Humble Is Streamlining Its Retail Credit Operations." National Petroleum News (February 1960): 124-126.

"API Chief Discusses Intangible Drilling Costs." Public Utilities Fortnightly 78 (September 15, 1966): 52, 54-56.

"Sunny Side Up: Critics Say Occidental Petroleum Corporation Trumpets Good Tidings, Plays Down Rest; Rosy Earnings Reports Don't Spell Out Special Items; Wired in to Wall Streeters." Wall Street Journal, 27 November 1967, 1.

"Accountants to Weigh Standardizing Reports of Oil, Mine Concerns; Current Variations Criticized As Clouding Investors' View of Individual Firm's Performance." Wall Street Journal, 10 November 1969, 14.

Field, Robert E. "Excerpts from Accounting Research Study No. 11: Financial Reporting in the Extractive Industries." Journal of Accountancy 128 (November 1969): 65-67.

"Wildcatter Wrangle: Accountants Striking Dry Well in Attempts to Significantly Change Oil-Company Rules." Wall Street Journal, 16 February 1972, 32.

"Pouring Oil on Troubled Auditors." Accountant (Eng.) 166 (February 1972): 181.

"Spread of 'Full Cost' of Oil-Gas Accounting to Other Industries Is Feared by Expert." Wall Street Journal, 28 April 1972, 4.

DRILLING FUNDS, PROGRAMS, JOINT VENTURES, AND PARTNERSHIPS

"SEC Issued Guidelines on Registration Statements by Some Oil, Gas Drilling Programs." Wall Street Journal, 21 January 1970, 4.

"A New Twist in Oil and Gas Drilling Ventures Is Stirring Controversy." Wall Street Journal, 1 September 1971, 1.

"Oil and Gas Drilling Funds Filed With SEC Fell to $503 Million in Offerings for First Half from $1.9 Billion for 1970 Period." Wall Street Journal, 1 September 1971, 1.

"Down the Hole: Small Investors Lured into Oil Drilling Funds Often Miss the Breaks; but They Still Flock to Funds, and So Do Regulators; a 'Little Fish' Goes Under." Wall Street Journal, 18 February 1972, 1.

"Watch Out for 'Conflicts of Interest' Oil and Gas Drilling Investors Are Warned." Wall Street Journal, 29 March 1972, 1.

"SEC Tightened Rules for Filing Exemptions of Small Drilling Funds; Oil and Gas Programs Must Limit Advertising to Get Exception, but Cutoff Size Is Increased." Wall Street Journal, 11 October 1972, 2.

ELECTRONIC DATA PROCESSING

Holland, Barney B. "Accounting System Helps Distributor Open Fuel Oil Market." Journal of Machine Accounting, Data Processing, Systems and Management 15 (July 1964): 4-6.

AUDITING

"Proposed FPC Regulations Specify Independent Audit." Journal of Accountancy 113 (May 1962): 18.

"FPC Extends Responsibility of Independent Auditors." Journal of Accountancy 125 (February 1968): 14.

PIPELINING AND TRANSMISSION

"Proposed FPC Regulations Specify Independent Audit." Journal of Accountancy 113 (May 1962): 18.

"FPC Extends Responsibility of Independent Auditors." Journal of Accountancy 125 (February 1968): 14.

"O'Connor Urges Accountants to Help FPC Develop Rules." Journal of Accountancy 125 (May 1968): 12, 14.

"AICPA Committee Responds to FPC Reporting Proposal." Journal of Accountancy 127 (February 1969): 12.

"FPC Seeks Accountants' Aid in Rate-Making Procedures." Journal of Accountancy 127 (February 1969): 10, 12.

"FPC Chief Accountant Issues New Accounting Interpretation." Journal of Accountancy 128 (December 1969): 10-12.

"Cost of All Gas Drilling Could Be Passed On Under FPC Proposal." Wall Street Journal, 6 October 1970, 4.

"FPC Eased Its Rule for Pipeline Firms' Accounting of Costs; Write-Offs for Gas Exploration, Unsuccessful Development Are Put on a More Favorable Basis." Wall Street Journal 8 November 1971, 10.

"FPC Announced Accounting Change That May Curtail Interstate Pipeline Companies Use of Advance Payments for Natural Gas Exploration." Wall Street Journal, 11 November 1971, 3.

"FPC Proposed Requiring Natural Gas Pipeline Firms to Furnish Agency Annually With Independent Appraisal of Their Recoverable Gas Reserves." Wall Street Journal, 6 January 1972, 16.

"Arthur Young & Co. Gets Natural Gas Survey Job." Journal of Accountancy 133 (June 1972): 17.

"FPC Moved to Required Interstate Pipeline Companies to Submit Detailed Data on Their Estimated Gas Reserves in Annual Reports Filed with Commission." Wall Street Journal, 1 November 1972, 4.

FUTURES CONTRACTS

"American Petroleum Institute Discussed Possibility of Futures in Gasoline, Heating Oil." Wall Street Journal, 14 November 1962, 26.

"Commodity Exchange Mulling Possible Trading in Propane Gas Futures." Wall Street Journal, 18 January 1967, 28.

"Trading in Liquefied Petroleum Gas Futures at Commodity Exchange, Inc. to Start Shortly After Labor Day." Wall Street Journal, 16 June 1967, 22.

"Trading in Propane Futures Opened Sept. 18 on New York Commodity Exchange." Wall Street Journal, 19 September 1967, 28.

OIL AND GAS
BOOKS
1960-1972

Alfred, John Burnis. "Accounting for the Oil Producing Industry," in J. K. Lasser's Handbook of Accounting Methods, 3rd ed. 1964, 720-733.

American Gas Association. Internal Auditing Procedures, Prepared by the Internal Auditing Committees of the American Gas Association and the Edison Electric Institute. Arlington, Va., 1972.

American Petroleum Institute. Estimation and Valuation of Underground Oil and Gas Reserves, 1971.

Crandall, John R.; Glanville, J. W.; and Cookenboo, L. "Cost of Acquiring and Operating Mineral Properties; Petroleum and Natural Gas," in Robie, Edward H., ed. Economics of the Mineral Industries, 2nd ed. New York: American Institute of Mining, Metallurgical, and Petroleum Engineers, 1964, 223-243.

Davis, Pinkney L. "Case Studies of Responsibility Accounting and Reports," in National Conference of Electric and Gas Utility Accountants. New York: American Gas Association and Edison Electric Institute, 1961, F4-43.

Dodson, C. R. "Bank Loans on Oil and Gas Production," in Robie, Edward H., ed. Economics of the Mineral Industries, 2nd ed. New York: American Institute of Mining, Metallurgical, and Petroleum Engineers, 1964, 275-279.

Grayson, C. Jackson, Jr. Decisions Under Uncertainty; Drilling Decisions by Oil and Gas Operators. Boston: Harvard University, Graduate School of Business Administration, 1960.

Griffin, Charles H. "Multiple Products Costing in Petroleum Refining," in Solomons, David, ed. Studies in Cost Analysis. 1968, 182-191. [reprinted from Journal of Accountancy 105 (March 1958)]: 46-52.

Hughes, Richard V. Oil Property Valuation. New York: John Wiley and Sons, 1967.

Legion, Walter A. "Special Features of Petroleum Accounting," in Robie, Edward H., ed. Economics of the Mineral Industries, 2nd ed. New York: American Institute of Mining, Metallurgical, and Petroleum Engineers, 1964, 450-462.

Lovejoy, Wallace F., and Homan, Paul T. Methods of Estimating Reserves of Crude Oil, Natural Gas, & Natural Gas Liquids. Resources for the Future Series. Baltimore: Johns Hopkins University Press, 1965.

Matthews, M. W. "Operating Methods in the Oil and Gas Industry," in Oil and Gas Production and Taxes. Toronto: Canadian Tax Foundation, 1963, 92-135.

Canadian Tax Foundation. Oil and Gas Production and Taxes. Toronto: Canadian Tax Foundation, 1963.

Porter, Stanley P. Petroleum Accounting Practices. New York: McGraw-Hill, 1965.

Simpson, R. A. "Canadian Petroleum Industry," in Canadian Tax Foundation, Oil and Gas Production and Taxes. Toronto: Canadian Tax Foundation, 1963, 1-91.

Tilly, Virgil S., Jr. "Audit of an Oil Operator," in Palen, Jennie M., ed. Encyclopedia of Auditing Techniques, 1966, 995-1039.

Toussaint, J. R. "Drilling Contractors," in Encyclopedia of Cost Accounting Systems. Englewood Cliffs, N.J.: Prentice-Hall, 1965, 253-290.

Usry, Milton F. Capital-Expenditure Planning and Control. Austin: University of Texas, Bureau of Business Research, 1966.

OIL AND GAS
BOOKLETS, REPORTS, RESEARCH STUDIES
1960-1972

Alfredson, F. K. *Study of Some Accounting Problems in the Oil Industry*. St. Lucia, Australia: University of Queensland Press, August 14, 1964.

American Petroleum Institute. Division of Finance and Accounting. *Report of Certain Petroleum Industry Accounting Practices*. New York, 1965.

American Petroleum Institute. Division of Finance and Accounting. *Report on Certain Petroleum Industry Accounting Practices, Resurvey as of December 31, 1966*. New York, 1967.

Andersen, R. Clifton, and Hewitt, Charles M. *Management and Financial Controls for Oil Jobbers*. Bloomington, Indiana: Indiana University, August 1961.

Arthur Andersen and Company. *Accounting for Oil and Gas Exploration Costs; for Purposes of Study and Discussion*. Chicago, January 1963.

Arthur Andersen and Company. *FPC Rule Making on Accounting for Gas and Oil Exploration and Development Costs*. Cases in Public Accounting Practice, Vol. 9. Chicago, 1972.

Arthur Andersen and Company. "Intangible Development Costs Applicable to Productive Oil and Gas Properties," in *Accounting and Reporting Problems of the Accounting Profession*. September 1960, 85-87.

Arthur Andersen and Company. "Oil and Gas Exploration Costs," in *Accounting and Reporting Problems of the Accounting Profession*. September 1960, 89-94.

Arthur Andersen and Company. "Producing Oil and Gas Working Interests Purchased Subject to Retained Production Payments," in *Accounting and Reporting Problems of the Accounting Profession*. September 1960, 95-99.

Coutts, W. B. *Accounting Problems in the Oil and Gas Industry*. Toronto: Canadian Institute of Chartered Accountants, Committee on Accounting and Auditing Research, 1963.

North American Securities Administrators Association. Committee on Oil and Gas Interests. *Guidelines for the Registration of Oil and Gas Drilling Programs*. Blue Sky Reports, No. 422. Chicago: Commerce Clearing House, 1971.

Porter, Stanley P. Full Cost Accounting: The Problem it Poses for the Extractive Industries. New York: Arthur Young & Company, 1972.

Price Waterhouse & Company. Survey of Financial Reporting and Accounting Developments in the Petroleum Industry Based on 1968 Reporting. New York, December 1969.

Price Waterhouse & Company. Survey of Financial Reporting and Accounting Developments in the Petroleum Industry, Based on 1969 Reporting. New York, 1970.

Price Waterhouse & Company. Survey of Financial Reporting and Accounting Developments in the Petroleum Industry Based on 1970 Reporting. New York, 1971.

Price Waterhouse & Company. Survey of Financial Reporting and Accounting Developments in the Petroleum Industry Based on 1971 Reporting. New York, 1972.

Sparling, Richard C.; Anderson, Norma J.; Emerson, John D.; and Winger, John G. 1964 Financial Analysis of Thirty-One Petroleum Companies. New York: Chase Manhattan Bank, June 1965.

OIL AND GAS
ARTICLES
1960-1972

GENERAL

Abel, T. P. "Underground Natural Gas Storage Accounting." *Canadian Chartered Accountant* 80 (May 1962): 471-477.

Altman, Richard B., and Lindberg, Charles S. "Oil and Gas: Nonoperating Oil and Gas Interests' Liability for Post-Production Costs and Expenses." *Oklahoma Law Review* 25 (November 1972): 572-577.

Bee, J. Kevin. "Some Problems in Oil Exploration Accounting." *Australian Accountant* 34 (June 1964): 307-313.

Bowles, John T.; Rooney, John M.; and Waller, Robert E. "Full Costing for Petroleum Exploration." *Canadian Chartered Accountant* 99 (December 1971): 421-428.

Cole, James F. "Rates of Return and Full Cost Accounting in the Oil Industry." *Canadian Chartered Accountant* 89 (September 1966): 202-204.

Coutts, W. B. "Accounting Problems in the Oil Industry." *Canadian Chartered Accountant* 79 (October 1961): 394-398.

Demers, Donald P. "Offshore Contract Drilling Industry." *Financial Analysts Journal* 26 (January/February 1970): 32-44.

Detlefsen, William K. "Full Cost Accounting in the Oil and Gas Industry." *Canadian Chartered Accountant* 92 (April 1968): 273-276.

Field, Robert E. "Extractive Industries--Applicable Accounting Concepts and Principles." *Price Waterhouse Review* 15 (Spring 1970): 16-29.

Finnell, Jack C.; Ayer, Leland G.; and Harris, Frank B. "Full Costing in the Oil and Gas Producing Industry." *Management Accounting* 48 (January 1967): 47-52.

Foster, George. "ARS No. 11, Historical Costs Without Perspective." *Abacus* (Australia) 7 (June 1971): 73-84.

Glezen, G. William, Jr. "Need for Accounting Postulates: ARS 11 as a Case in Point." Louisiana Certified Public Accountant 30 (Fall 1970): 40-43.

Harding, S. R. "Accountants in the Oil Industry." Accountant (Eng.) 151 (November 7, 1964): 567-569.

Harnden, Willard J., and Rosenbaum, Allan S. "Balance Sheet Presentation of Notes Payable." Colorado CPA Report 31 (January 1967): 10-12. Oklahoma CPA 6 (July 1967): 32-34.

Harpster, Wayne W. "Total Cost Accounting for Petroleum Exploration Costs." Management Controls 12 (August 1965): 159-161.

Hazelwood, C. T. "Accounting for Exploration and Related Expenditure." Australian Accountant 41 (April 1971): 137-138.

Huie, William O. "Apportionment of Oil and Gas Royalties." Harvard Law Review 78 (April 1965): 1113-1145 and (June 1965): 1534-1567.

"Industry Approach to Improved Financial Reporting." Journal of Accountancy 122 (September 1966): 35-36.

Johnson, Leonard W. "Offshore Drilling and Equipment Industry." Financial Analysts Journal 24 (March/April 1968): 58-60.

Johnson, Robert T. "Full Cost vs. Conventional Accounting in the Petroleum Industry." CPA Journal 42 (June 1972): 479-484.

Klingstedt, John P. "Effects of Full Costing in the Petroleum Industry." Financial Analysts Journal 26 (September/October 1970): 79-86.

Klingstedt, John P. "Accounting Research Study No. 11--A Step in the Right Direction." Texas CPA 43 (July 1970): 46-48.

Koons, R. L. "More on Accounting for Oil Prospecting Costs." CPA Journal 42 (December 1972): 989.

Leonard, E. C. "Problems of Difference and Conformity in Reporting Practices of Oil Companies." Oklahoma CPA 2 (October 1963): 12-17.

Lourens, Roy. "Development of Reporting Standards for the Extractive Industries." Australian Accountant 42 (September 1972): 329-334.

Mulcahy, Gertrude, ed. "Financial Reporting in the Extractive Industries: AICPA Accounting Research Study No. 11." Canadian Chartered Accountant 96 (June 1970): 421-423 and 97 (July 1970): 45-47.

"Oil Company Accounting." Accountancy (Eng.) 82 (July 1971): 404-408.

Patz, Dennis H., and Boatzman, James R. "Accounting Principles Formulation in an Efficient Market Environment." Journal of Accounting Research 10 (Autumn 1972): 392-403.

"Pension Plans of the Petroleum Industry." Pension and Welfare News 6 (January 1970): 53-54.

Reed, Randal W. "Petroleum Industry Accounting in Perspective." Financial Executive 38 (April 1970): 66-70.

Rice, Carroll A. "Total Cost Accounting." Colorado CPA Report 34 (Spring 1970): 11-19.

Smith, C. Aubrey. "Generally Accepted Accounting Principles and the Petroleum Industry." Texas Certified Public Accountant 37 (July 1964): 3-10.

Stewart, Charles A. "Oil Industry Accounting Practices; A Brief Description of Oil Industry Accounting for the Businessman." Price Waterhouse Review 8 (Spring 1963): 28-32.

Stevens, Ross. "Do Your Accounts Tell All?" Fueloil and Oil Heat 19 (March 1960): 77-80, 154.

Stormonth, S. G. "Petroleum Exploration Accounting." Chartered Accountant in Australia 36 (May 1966): 820-832.

Stow, Louis J. "Accounting Research Study No. 11 As It Relates to the Oil and Gas Industry." Texas CPA (July 1970): 48-51, 67.

Surratt, Marshall E. "Oil Well Drilling Contractors--A Statement Study." Bulletin of the Robert Morris Associates 44 (November 1961): 73-80.

Thomas, Wayne W. "Appraising Profit in the Crude Oil Producing Industry." NAA Bulletin 43 (December 1961): 61-68.

Thomson, John W. "Accounting for Oil Drilling Operations." Canadian Chartered Accountant 78 (February 1961): 165-171.

Weathers, Henry T. "Eclectic Accounting Approach." Management Accounting 54 (December 1972): 36-38.

DISCLOSURES AND RESERVE RECOGNITION ACCOUNTING

Johnson, Robert T. "Examination of Oil and Gas Reserves." Journal of Accountancy 134 (November 1972): 79-82.

DRILLING FUNDS, PROGRAMS, JOINT VENTURES, AND PARTNERSHIPS

Allan, Edward B. "Joint Venture Accounting in the Oil Industry." Canadian Chartered Accountant 77 (November 1960): 406-411.

Hershman, Arlene. "Drilling Funds: What Risk? What Reward?" Dun's Review 99 (June 1972): 39-41, 116, 118, 120.

McIntosh, Donald L., and Joseph, Gifford E. "Accounting and Tax Aspects of Oil and Gas Joint Ventures." Ohio State Law Journal 26 (Fall 1965): 546-566.

Nielsen, Carl C. "Reporting Joint-Venture Corporations." Accounting Review 40 (October 1965): 795-804.

Raub, Elizabeth B. "Accounting Procedures for Joint Venture Operations in the Oil Industry." Woman CPA 29 (October 1967): 5-6, 10, 13.

Sanders, David T., and Serio, S. Barnett. "Financial Analysis of Oil and Gas Drilling Programs." Trusts and Estates 111 (March 1972): 186-189.

Stephens, Lowell W. "Guidelines for the Registration of Oil and Gas Drilling Programs." Texas CPA 44 (April 1972): 51-53.

Sullivan, Joseph P. "Oil and Gas Investment Programs: A Brief Survey." Business Lawyer 26 (April 1971): 1027-1038.

Thomas, Thad W. "Real Estate Investor's Guide to Oil and Gas Drilling Programs." Real Estate Review 1 (Spring 1971): 84-93.

ECONOMICS, FINANCE, AND EVALUATION

Christian, J. Roy. "Oil Properties in Trust--Profitable Problem Solving." Trusts and Estates, 3 (June 1972): 454-457, 511-512.

Christopher, David W., and Henderson, John G." 'Total Energy' in the Gas Industry." Price Waterhouse Review 11 (Winter 1966): 42-47.

Clark, Robert L. "Energy Industry in the World of Tomorrow." Financial Analysts Journal 22 (September/October 1966): 59-66.

Clifford, W. Kelly. "Investing in Oil and Gas." Arthur Andersen Chronicle (June 1969): 24-28.

Fulgham, J. Rawles, Jr. "Recent Trends in Financing the Petroleum Industry." Bulletin of the Robert Morris Associates 46 (October 1963): 59-62.

Grady, Allan B. "Dealing With A Declining Asset: Evaluating Fractional Interests in Oil and Gas Wells." Trusts and Estates 906 (August 1967): 708, 710-712.

Guinness, Robert C. "Economics of Energy." Financial Analysts Journal 24 (January/February 1968): 47-50.

Hershman, Arlene. "Drilling Funds: What Risk? What Reward?" Dun's Review 99 (June 1972): 39-41, 116, 118, 120.

Ireton, Greg. "Should Commerical Banks Make Oil Loans?" Bulletin of the Robert Morris Associates 49 (October 1966): 14-22.

McGregor, D. A. "Mergers in the Mining and Oil Industries." Canadian Chartered Accountant 76 (February 1960): 179-182.

Reso, Jerome J., Jr. "Simplified Explanation of the ABC Transaction." Loyola Law Review 10 (1960-1961, No. 2): 249-258.

Sanders, David T., and Serio, S. Barnett. "Financial Analysis of Oil and Gas Drilling Programs." Trusts and Estates 111 (March 1972): 186-189.

Spriggs, Dillard. "Forecasting Oil Industry Profits." Financial Analysts Journal 22 (September/October 1966): 49-52.

Stephens, Lowell W. "Guidelines for the Registration of Oil and Gas Drilling Programs." Texas CPA 44 (April 1972): 51-53.

Sullivan, Joseph P. "Oil and Gas Investment Programs: A Brief Survey." _Business Lawyer_ 26 (April 1971): 1027-1038.

Thomas, Thad W. "Real Estate Investor's Guide to Oil and Gas Drilling Programs." _Real Estate Review_ 1 (Spring 1971): 84-93.

PLANNING, CONTROL, AND BUDGETING

Barclay, I. T. "Extension of Variance Analysis." _Cost Accountant_ (Eng.) 41 (August 1963): 305-307.

Deam, R. J. "Linear Programming in Management Accounting." _Accountant_ (Eng.) 157 (September 23, 1967): 387-391.

Deam, R. J. "Linear Programming in Management Accounting." _Management Accountant_ (India) 3 (February 1968): 63-73. (reprinted from _Accountant_ (Eng.) 157 (September 23, 1967): 387-391.

Devine, M. D., and Lesso, W. G. "Models for the Minimum Cost Development of Offshore Oil Fields," _Management Science_ 18 (April 1972): 378-387.

Friend, B. E., and Phillips, K. C. "Use of Current Computer Concepts in Inventory Management and Other Management Control Techniques." _Accountancy_ (Eng.) 77 (December 1966): 846-854.

Gandek, B. S. "Reporting Supply and Distribution Terminal Costs to Operating Management." _Cost and Management_ (Canada) 39 (January 1965): 3-6.

Jones, Donald P. "Budgetary Control in the Sun Oil Company." _Federal Accountant_ 11 (March 1962): 57-66.

Lafferty, John V. "Control Report Package for an Oil Producing Company." _NAA Bulletin_ 45 (November 1963): 55-61.

Most, Kenneth S. "The Cost Center Problem in the Oil Industry." _Management Accounting_ 54 (December 1972): 39-41, 47.

Mukherjee, A. C. "Control Over Stock Loss in Petroleum Industries." _Management Accountant_ (India) 2 (May 1967): 229-234.

Prouty, M. D. Jr., and Severin, J. R., Jr. "Serving Two Masters With One Accounting System." NAA Bulletin 41 (April 1960): 49-58.

Robb, Thomas C. "Exploration Department of an Integrated Oil Company." Arthur Young Journal 8 (July 1960): 20-32.

Vasinathan, J. "Choice of Operating Alternatives With the Aid of Cost Analysis." Industrial Accountant 5 (October 1965-March 1966): 13-16.

"What Are the Present Practices in Truck Metering and Ticketing?" Fueloil and Oil Heat 19 (March 1960): 12-13.

ELECTRONIC DATA PROCESSING

Brook, C. N. "Computer Control for Natural Gas Conversion." Management Accounting (Eng.) 49 (February 1971): 51-56.

Elliott, Norman J. "Electronic Processing of Oil and Gas Distributions." Journal of Accountancy 117 (March 1964): 79-81.

Friend, B. E., and Phillips, K. C. "Use of Current Computer Concepts in Inventory Management and Other Management Control Techniques." Accountancy (Eng.) 77 (December 1966): 846-854.

Harper, Joseph H., and McAlpine, Lloyd R. "EDP in Management: A Study in the Oil Industry." Canadian Chartered Accountant 88 (January 1966): 43-47.

Holmes, Donald C. "Computers in Oil 1967-1987." Quarterly (Touche Ross) 13 (September 1967): 16-23.

Milner, Roy W. "Optical Card Punch Speeds Credit Card Accounting." Journal of Data Management 4 (June 1966): 78-81.

AUDITING

Anderson, H. F. "Preventive Auditing." Internal Auditor 17 (March 1960): 21-27.

Miles, Francis L., and Glezen, G. William, Jr. "Auditing the Cost of Finding Oil and Gas." Louisiana Certified Public Accountant 26 (October 1966): 18-25, 27-29.

REFINING

Barkman, Arnold I. "Management Information in an Oil Refinery." Office 71 (March 1970): 12-15.

Berman, Norman D. "Profit Analysis Practices in an Oil Refining Company." NAA Bullentin 42 (July 1961): 63-68.

Dutt, A. K. "Costing for Petroleum Refinery." Chartered Accountant (India) 12 (May 1964): 588-601.

Fox, John L. "Accounting for Petroleum Refiners; A Brief Guide to Cost Accounting, Inventory Valuation and Inventory Control in Petroleum Refining." Price Waterhouse Review 9 (Summer 1964): 42-50.

Gibson, Ray G. "Refining and Marketing Margins for Domestic Gasoline." Financial Analysts Journal 28 (January/February 1972): 38-42, 88-90.

Gregorio, D. V., Jr. "Overview of Oil Refinery Accounting." SGV Journal (No. 2, 1968): 8-10.

Griffin, James M. "Process Analysis Alternative to Statistical Cost Function: An Application to Petroleum Refining." American Economic Review 62 (March 1972): 46-56.

Points, A. J. "Multiple-Product Operating Statements in an Oil Refinery." NAA Bulletin 43 (June 1962): 85-92.

Said, Omar H. M. "Ways to Joint Product Costing of Refined Oil Stocks." Singapore Accountant 6 (1971): 78-83.

PIPELINING AND TRANSMISSION

Bercovici, Alex. "Process Control and Computer Application in the Gas Transmission Industry." Canadian Chartered Accountant 93 (August 1968): 88-92.

Brock, Horace R. "Joint-Cost Allocation--Not a Rate Basis." NAA Bulletin 44 (February 1963): 19-26.

Litke, Arthur L. "Different Masks of Regulatory Accounting." Public Utilities Fortnightly 78 (September 1, 1966): 28-38.

Litke, Arthur L. "Impact of FPC Reporting and Accounting Changes." Public Utilities Fortnightly 89 (January 20, 1972): 15-21.

Nissel, Hans E. "Impact of Cost Allocations Upon Future of the Natural Gas Industry." Public Utilities Fortnightly 66 (October 13, 1960): 512-524.

Rosenberg, Joel B. "Cost Study of the Natural Gas Distribution Industry." Public Utilities Fortnightly 90 (October 1972): 41-43.

Thompson, Russell G.; Proctor, Michael S.; and Hocking, Ronald R. "Investment-Borrowing Decisions in Natural Gas Transmissions." Management Science 18 (June 1972): 544-554.

Thornton, Fred A. "FPC Allocation of Gas Storage Costs and Pricing." Public Utilities Fortnightly 89 (May 25, 1972): 11-15.

Unnevehr, Clarence. "Round and Round with Gas Cost Allocations." Public Utilities Fortnightly 73 (January 16, 1964): 19-25.

OIL AND GAS DISSERTATIONS 1960-1972

Allain, Louis John Joseph. "Capital Investment Models of the Oil and Gas Industry: A Systems Approach." Ph.D. dissertation, Purdue University, 1970.

Bentley, James Richard. "An Examination of Accounting for Depletion in the Petroleum Industry." D.B.A. dissertation, University of Washington, 1968.

Culpepper, Robert Carlton. "A Study of the Effect of Certain Alternative Accounting Methods of Regulatory Decisions in the Gas Utility Industry." Ph.D. dissertation, University of Arkansas, 1969.

Duckworth, Patricia Lynch. "Accounting for Preproduction Costs in the Oil and Gas Industry." D.B.A. dissertation, University of Colorado, 1966.

Fetzer, Joseph Baer. "An Analysis of the Applicability of Certain Optimal Capital Structure Concepts to a Selected Group of Domestic Integrated Oil Companies." Ph.D. dissertation, Stanford University, 1964.

Gerwig, Robert William. "Natural Gas Production: A Study of Costs of Regulation." Ph.D. dissertation, University of Chicago, 1963.

Hansen, Robert Eugene. "An Historical Development of Accounting for Joint Ventures." D.B.A. dissertation, Indiana University, 1968.

House, William Clyde, Jr. "The Effects of Errors in Estimation Upon Measures of Return Used by Managements of Oil and Gas Pipeline Companies to Make Investment Decisions." Ph.D. dissertation, The University of Texas at Austin, 1965.

Kaufman, Gordon Mayer. "Statistical Decision and Related Techniques in Oil and Gas Exploration." Ph.D. dissertation, Harvard University, 1961.

Klingstedt, John Paul. "Full Costing in the Petroleum Industry and Its Implications for Accounting Principles and Practices." Ph.D. dissertation, North Texas State University, 1969.

Lehman, Edward Richard. "Profits, Profitability, and the Oil Industry." Ph.D. dissertation, New York University, 1964.

Maass, Randal Oscar. "Empirical Measurements of the Cost of Capital for Selected Companies in the Petroleum Industry." D.B.A. dissertation, University of Southern California, 1962.

McArthur, John Hector. "A Study of Debt Financing and the Formulation of Debt Policy in the Pipeline Sector of the Oil Industry Between 1941 and 1960." Ph.D. dissertation, Harvard University, 1963.

Melody, William Harry. "Public Utility Regulation: Economic Analysis, Accounting Control, and Federal Power Commission Rate Policy." Ph.D. dissertation, The University of Nebraska--Lincoln, 1966.

Mohamed, Ahmed Sorour. "Scheduling of Production in the Petroleum Refining Industry: An Operations Research Study." Ph.D. dissertation, University of Pittsburgh, 1961.

Posey, Joshua Marion. "An Inquiry into the Economic, Financing, and Accounting Factors Involved in Expansion of Natural Gas Utility Distribution Systems," Ph.D. dissertation, University of Arkansas, 1968.

Schmeltz, William Frederick. "Accounting and Management Control Practices in Petroleum Refining." Ph.D. dissertation, Case Western Reserve University, 1966.

Sellers, David Henry Aikins. "Investment Planning Under Uncertainty for Oil and Gas Producing." Ph.D. dissertation, The Pennsylvania State University, 1965.

Smith, Alan Frederick. "Valuation and Its Possible Application and Presentation in Accounting With Special Reference to the Production of Oil." D.B.A. dissertation, University of Colorado, 1968.

Smith, Langford Wheaton, Jr. "An Approach to Costing Joint Production Based on Mathematical Programming With An Example from Petroleum Refining." Ph.D. dissertation, Stanford University, 1962.

Taker, Abdul Hady Hassan. "Income Determination in the International Petroleum Industry." Ph.D. dissertation, University of California, Berkeley, 1964.

Thornton, Fred Arnold. "An Examination of the Economic Implications Inherent in Federal Power Commission Cost Allocation Policies for Natural Gas Pipeline Companies for the Period 1952-1967." Ph.D. dissertation, The Ohio State University, 1970.

1960-1972

Mining

MINING
1960-1972

GENERAL

Adamson, John Q. "Accounting for Sulphur Mining," in J. K. Lasser's Handbook of Accounting Methods, 3rd ed. 1964, 892-909.

Canadian Institute of Chartered Accountants. Committee on Accounting and Auditing Research. Financial Reporting for Non-Producing Mining Companies. Toronto, 1967.

Corcoran, Eileen T. "Current Studies and Concepts," Woman CPA 32 (September 1970): 11-12.

Coutts, W. B. "Mining Accounts." Canadian Chartered Accountant 83 (July 1963): 55-57.

Davis, John L. "Accounting for Electrolytic Refining Copper." Management Accounting 50 (October 1968): 49-52.

Davis, Robert H. "Production Costing in Open Pit Mining." Management Accounting 52 (January 1971): 39-41.

Emblen, D. J. "Mine Accounting," in Handbook for Small Mining Enterprises in Montana. Butte, Montana: Montana Bureau of Mines and Geology, May 1964, 117-125.

Fagerberg, Dixon, Jr. "Analyzing Results of Mining Enterprises." Management Controls 12 (August 1965): 162-167.

Foster, George J. "Mining Inventories in a Current Price Accounting System." Abacus (Australia) 5 (December 1969): 99-118.

Getliffe, D. B. "Brief History of the Treatment of Capital Expenditure in the Accounts of Gold Mining Companies." South African Accountant 8 (March 1961): 8-11.

Griswold, George G., Jr. "Minerals Beneficiation Accounting," in Handbook for Small Mining Enterprises in Montana. Butte, Montana: Montana Bureau of Mines and Geology, May 1961, 126-180.

Hooper, Dudley W. "Coal Board Story; Integrated E.D.P. in a Nationalized Industry." Accountant (England) 147 (December 5, 1962); 764-768 and (December 22, 1962): 802-805.

Jones, William C. "Coal Mines," in Encyclopedia of Cost Accounting Systems. Englewood Cliffs, N.J.: Prentice-Hall, 1965, 167-195.

Leister, Mortimer A. "Accounting for Coal Mining Companies of Medium Size," in J. K. Lasser's Handbook of Accounting Methods, 3rd ed., 1964, 253-269.

Little, A. J., and MacDonald, W. E. Mining Accounting in Canada; An Outline of Accounting and Income Tax Problems Peculiar to the Industry. Toronto: Canadian Institute of Chartered Accountants, 1960.

Mathur, A. C. "Costing in Gypsum Mining Industry," Management Accountant (India) 1 (March 1966): 125-129.

McCarthy, Robert M. "Financial and Tax Accounting for Production Payments." Pennsylvania CPA Spokesman 38 (August 1967): 5, 8-9.

McQuail, James A. "Accounting Principles and Practices." Pennsylvania CPA Spokesman 38 (August 1967); 4, 10.

Montana Bureau of Mines and Geology. Handbook for Small Mining Enterprises in Montana. Butte, Montana, May 1964.

Nethercott, L. J. "Financial Statements of Extractive Industries." Australian Accountant 40 (October 1970): 433-435.

Peloubet, Maurice E. "Accounting for the Extractive Industries," in Robie, Edward H., ed. Economics of the Mineral Industries, 2nd ed. New York: American Institute of Mining, Metallurgical, and Petroleum Engineers, 1964, 403-450.

Pillai, G.G. "Some Distinctive Features in Mine Accounting Practice and Valuation." Management Accountant (India) 3 (November 1968): 562-569.

Robertson, A. S. "Uses of Gold Mining Companies' Accounts and Reports." South African Accountant 7 (June 1960): 56-59.

Stout, K. S., and Emblen, D. J. "Production Costs," in Handbook for Small Mining Enterprises in Montana. Butte, Montana: Montana Bureau of Mines and Geology, May 1964, 99-104.

Tyler, Ottis Jan. "Sale or Lease of Mineral Interests--Fact or Fiction." Southwestern Law Journal 17 (December 1963): 567-590.

Wilcox, F. "Determining the Book Value of Mining Rights." Canadian Chartered Accountant 85 (October 1964): 262-266.

JOINT VENTURES

Robberson, R. T. "Mining Joint Operating Agreements--No Longer Hen's Teeth," in Rocky Mountain Mineral Law Institute, 14th, 1968 Proceedings. New York: Matthew Bender & Company, 1968, 309-381.

Robberson, R. T., and Van Bebber, William G. "Whatever Happened to the Model Form Mining Joint Operating Agreement?" in Rocky Mountain Mineral Law Institute, 17th, 1971 Proceedings. New York: Matthew Bender & Company, 1972, 407-433.

Williams, J. Harley, Jr. "Unitization of Mining Properties," in Rocky Mountain Mineral Law Institute, 17th, 1971 Proceedings. New York: Matthew Bender & Company, 1972, 245-279.

ECONOMICS, FINANCE, AND EVALUATION

Bloomenthal, Harold S. "Financing Mineral Development," in Rocky Mountain Mineral Law Institute, 13th, 1967 Proceedings. New York: Matthew Bender & Company, 1967, 115-142.

Boericke, William F., and Bailey, Carroll C. "Mine Financing," in Robie, Edward H., ed. Economics of the Mineral Industries, 2nd ed. New York: American Institute of Mining, Metallurgical, and Petroleum Engineers, 1964, 245-275.

Bretey, Pierre R. "Bituminous Coal--Adequate Financing Essential for Future Growth." Financial Analysts Journal 24 (March/April 1968): 61-64.

Broadbent, Lorin J. "Eminent Domain Valuation of Land Containing Minerals." Appraisal Journal 29 (January 1961): 63-77.

Fisher, Franklin M.; Cootner, Paul H.; and Baily, Martin Neil. "Econometric Model of the World Copper Industry." Bell Journal of Economics and Management Science 3 (Autumn 1972); 568-611.

Higgins, W. Rodgers. "Valuation of Readily Marketable Inventories." Journal of Accountancy 118 (July 1964): 25-32.

Lemon, James D. "Problems in Mine Evaluation." Management Accounting 53 (September 1971): 46-48.

Matthews, A. Bruce. "Financial Projections to Determine Economic Feasibility of Proposed Mining Ventures." Arthur Andersen Chronicle 20 (July 1960): 233-244.

McSweeney, Thomas F. "Evaluation of Mineral Rights," in American Society of Appraisers, Appraisal and Valuation Manual 1960, 1960, 5, 153-160.

Olsen, John A., and Blaney, Terry A. "Copper Industry--Forecasting by Probabilities." Financial Analysts Journal 24 (March/April 1968): 36-40.

Pillai, G. G. "Some Distinctive Features in Mine Accounting Practice and Valuation." Management Accountant (India) 3 (November 1968): 562-569.

Randall, William J. "Problems in Lead Mining Appraisals." Appraisal Journal 28 (July 1960): 345-358.

Raymond, L. C. "Valuation of Mineral Property," in Robie, Edward H., ed. Economics of the Mineral Industries, 2nd ed. New York: American Institute of Mining, Metallurgical, and Petroleum Engineers, 1964, 131-165.

Vine, W. A. "Mine Financing," in Handbook for Small Mining Enterprises in Montana. Butte, Montana: Montana Bureau of Mines and Geology, May 1964, 194-212.

Way, Peter F. "Forecasting by Probabilities--Preface to Copper Industry Study." Financing Analysts Journal 24 (March/April 1968): 35.

Winters, Harry Joseph, Jr. "Financial Evaluation of Mineral Industry Capital Projects." Ph.D dissertation, University of Arizona, 1972.

News Releases and Short Comments

"Mining Stocks, Claims Soar as Timmers, Ont., Celebrates in Streets; Volume on Toronto Stock Exchange Broke Record Set in 1958 in Midst of Uranium Boom." Wall Street Journal, 17 April 1964, 2.

"Timmins, Ontario, Nickel Find Sparked Another Rush of Speculation on Toronto Stock Exchange." Wall Street Journal, 25 November 1964, 30.

"Commission Appointed by Ontario Government Will Begin Public Hearings February 25 on Stock-Price Gyrations of Windfall Oils and Mines, Ltd. and Other Mining Concerns in Province." Wall Street Journal. 19 February 1965, 9.

"Ontario's Securities Regulations for Mining Stock Sales Should Be Revamped, Committee Formed by Ontario Securities Commission Proposed." *Wall Street Journal*, 24 September 1968, 6.

"Speculative Mine Financing in Ontario Expected to Be Cut by New Securities Law That Cuts to 100% from 200% Maximum Markup That Can Be Charged by Underwriters of Unlisted Mining Shares Sold to Public." *Wall Street Journal*, 25 September 1968, 8.

REGULATION

News Releases and Short Comments

"Ontario's Securities Regulations for Mining Stock Sales Should Be Revamped, Committee Formed by Ontario Securities Commission Proposed." *Wall Street Journal*, 24 September 1968, 6.

"Speculative Mine Financing in Ontario Expected to Be Cut by New Securities Law That Cuts to 100% from 200% Maximum Markup That Can Be Charged by Underwriters of Unlisted Mining Shares Sold to Public." *Wall Street Journal*, 25 September 1968, 8.

BLACK LUNG

News Releases and Short Comments

"West Virginia Miners Accelerate Campaign Over 'Black Lung' Bill; Vexed by Weakened Senate One, They Threaten Long Walkout, Exodus; Plea Is Sent to President Nixon." *Wall Street Journal*, 7 March 1969, 2.

"Miners Union Gets Cold Reception from Militant 'Black Lung' Fighters." *Wall Street Journal*, 24 June 1969, 1.

"Tough Mine-Safety Bill Passed by Subcommittee of House Education & Labor Committee; Plan, Unlike Senate's Sets Payments for 'Black Lung' Victims: Dust Limit Is Rigid." *Wall Street Journal*, 7 August 1969, 15.

"Senate Rejected Special Assessment on Coal Industry; Substitute Plan Would Finance Health and Safety Research With Regular Treasury Funds." *Wall Street Journal*, 1 October 1969, 25.

"Senate Passed Unanimously Mine Safety Bill; Measure Limits Dust, Methane, Calls for Frequent Inspections, Aid to 'Black Lung' Victims." Wall Street Journal, 3 October 1969, p. 2.

"House Passed Bill Substantially Broadening Eligibility for Cash Benefits Under Government Program to Compensate Victims of 'Black Lung' Disease." Wall Street Journal, 11 May 1972, 11.

"Coal Mine Safety Bill Approved by Congress; Faces Possible Veto; Administration Objects to Costs of Compensation Proposal for Victims of 'Black Lung' Disease." Wall Street Journal, 19 December 1969, 7.

"President Nixon to Sign Mine Legislation Dec. 30; Mining Measure Includes Funds for Victims of 'Black Lung'; Cost Said to Worry President." Wall Street Journal, 30 December 1969, 6.

"Coal Mine Safety Bill Signed into Law; President Nixon Voices Reservations; Asserts Compensation Provision Is 'Limited, Unique,' to Expire Within Seven Years." Wall Street Journal, 31 December 1969, 2.

"Coalfield Clash: Bitter Miners Assert the 'Black Lung' Law is Filled with Loopholes; Disabled Men Say Benefits Are Difficult to Collect; 'Slipshod' Diagnoses Cited." Wall Street Journal, 24 September 1971, 1.

"Costly Black Lung Assistance Bill Causes Dilemma for Budget-Minded Republicans." Wall Street Journal, 10 May 1972, 4.

"House Passed Law Stiffening Health, Safety Rules Involving Coal Mines." Wall Street Journal, 30 October 1969, 33.

"Black Lung Benefits Act of 1972." Wall Street Journal, 21 June 1972, 16.

PLANNING, CONTROL, AND BUDGETING

Burke, Lamar. "Standard Costing in Kaolin Mining." Management Accounting 52 (May 1971): 22-24.

Durik, Emil J. "Cost Accounting and Control for Mining Companies." Management Accounting 53 (August 1971): 37-40.

Dutta, S. K. "Challenges of Cost Control in Coal Industry." Management Accountant (India) 6 (July 1971): 398-404.

Eshbach, Alton R. "Cost Reporting for Mobile Equipment in the Mining Industry." NAA Bulletin 42 (September 1960): 75-82.

Hooper, Dudley W. "Coal Board Story; Integrated E.D.P. in a Nationalized Industry." Accountant (England) 147 (December 5, 1962); 764-768 and (December 22, 1962): 802-805.

Jones, William C. "Coal Mines," in Encyclopedia of Cost Accounting Systems. Englewood Cliffs, N.J.: Prentice-Hall, 1965, 167-195.

Laughlin, B. "Mining Slant on Marginal Costs." Australian Accountant 38 (September 1968): 523-525.

Leonard, W. Drew. "Location, Development and Operation of an Iron Ore Mine." Financial Executive 32 (November 1964): 224-27, 30, 34.

Mathur, A. C. "Costing in Gypsum Mining Industry," Management Accountant (India) 1 (March 1966): 125-129.

McGhee, Archie. "Internal Auditing at Foote Mineral Company." Internal Auditor 18 (Winter 1961): 83-86.

Meyer, Manfred. "Applying Linear Programming to the Design of Ultimate Pit Limits." Management Science 16 (October 1969): 121-135.

Peterson, B. C.; Nelson, E. R.; and Beck, William A. U.S. Department of the Interior. Cost Accounting and Control at Calumet Division, Calumet and Hecla, Inc. Calumet, Michigan. Washington: Government Printing Office, 1962.

Stout, K. S., and Emblen, D. J. "Production Costs," in Handbook for Small Mining Enterprises in Montana. Butte, Montana: Montana Bureau of Mines and Geology, May 1964, 99-104.

Tyler, Paul M. "Cost of Acquiring and Operating Mineral Properties; Metal, Nonmetallic, and Coal," in Robie, Edward E., ed. Economics of the Mineral Industries, 2nd ed. New York: American Institute of Mining, Metallurgical, and Petroleum Engineers, 1964, 167-223.

AUDITING

Herreid, Thomas N. "Audit of a Coal Mining Company," in Palen, Jennie M., ed. Encyclopedia of Auditing Techniques. 1966, 393-427.

McGhee, Archie. "Internal Auditing at Foote Mineral Company." Internal Auditor 18 (Winter 1961): 83-86.

Nethercott, Leslie J. "Mineral Exploration Companies and the Auditor." Chartered Accountant in Australia 42 (November 1971): 4-7.

Pogson, Percy W. "Auditing--A Small Mining Company." Quarterly (Arizona Society of CPAs) 62 (November 1962): 13-14, 16, 18.

1973-1982

Oil & Gas

OIL AND GAS
PRONOUNCEMENTS AND REPORTS OF NONGOVERNMENTAL ACCOUNTING AND AUDITING STANDARDS BODIES
1973-1982

American Institute of Certified Public Accountants. Accounting Principles Board Committee on Extractive Industries. Accounting and Reporting Practices in the Oil and Gas Industry. New York, 1973.

American Institute of Certified Public Accountants. Auditing Standards Board. "Supplementary Information on the Effects of Changing Prices." "Supplementary Oil and Gas Reserve Quantity Information." Proposed Statements on Auditing Standards. New York, December 31, 1979.

American Institute of Certified Public Accountants. Auditing Standards Board. "Supplementary Information Required by the Financial Accounting Standards Board." Statement on Auditing Standards No. 27. New York, 1979.

American Institute of Certified Public Accountants. Auditing Standards Board. "Supplementary Oil and Gas Reserve Information." Statement on Auditing Standards No. 33. New York, October 1980.

American Institute of Certified Public Accountants. Auditing Standards Board. "Supplementary Mineral Reserve Information." Statement on Auditing Standards No. 40. New York, February 1982.

American Institute of Certified Public Accountants. Auditing Standards Board. "Using the Work of a Specialist." Statement on Auditing Standards No. 11. New York, 1975.

American Institute of Certified Public Accountants. Oil and Gas Committee, "Audits of Oil and Gas Producing Companies." Proposed Audit Guide. New York, 1982.

American Institute of Certified Public Accountants. Oil and Gas Reserve Data Committee. "Oil and Gas Reserve Information Required by Regulation S-X." Proposed Audit and Accounting Guide. New York, 1979.

Dyckman, Thomas. Effects of the Issuance of the Exposure Draft and FASB 19 on the Security Returns of Oil and Gas Producing Companies. Stamford, Conn.: Financial Accounting Standards Board, 1979.

Dyckman, Thomas R. Report 3, on the Effects of Statement No. 19, Financial Accounting and Reporting by Oil and Gas Companies, on Returns of Oil and Gas Company Securities. Stamford, Conn.: Financial Accounting Standards Board, March 1978.

Financial Accounting Standards Board. "Accounting by Oil and Gas Producing Companies for the Tax Benefits of Percentage Depletion." Proposed FASB Interpretation. Stamford, Conn., July 31, 1980.

Financial Accounting Standards Board. "Accounting for Exploratory Wells in Progress at the End of a Period." FASB Interpretation No. 36. Stamford, Conn., October 1981.

Financial Accounting Standards Board. "Accounting for Exploratory Wells in Progress at the End of a Period." Proposed FASB Interpretation. Stamford, Conn., November 26, 1980.

Financial Accounting Standards Board. "Accounting for Income Taxes--Oil and Gas Producing Companies." Statement of Financial Accounting Standards No. 9. Stamford, Conn., October 1975.

Financial Accounting Standards Board. "Accounting for the Effects of Certain Types of Regulation." Statement of Financial Accounting Standards No. 71. Stamford, Conn., 1982.

Financial Accounting Standards Board. "Analysis of Issues Related to Financial Accounting and Reporting in the Extractive Industries." FASB Discussion Memorandum. Stamford, Conn., December 23, 1976.

Financial Accounting Standards Board. "Applying FASB Statement No. 34 to Oil and Gas Producing Operations Accounted for by the Full Cost Method." FASB Interpretation No. 33. Stamford, Conn., August 1980.

Financial Accounting Standards Board. "Applying FASB Statement No. 34 to Oil and Gas Producing Operations Accounted for by the Full Cost Method." Proposed FASB Interpretation. Stamford, Conn., January 28, 1980.

Financial Accounting Standards Board. "Disclosures About Oil and Gas Producing Activities." Invitation to Comment. Stamford, Conn., May 13, 1981.

Financial Accounting Standards Board. "Disclosures About Oil and Gas Producing Activities." Proposed Statement of Financial Accounting Standards. Stamford, Conn., April 15, 1982.

Financial Accounting Stardards Board. "Disclosures About Oil and Gas Producing Activities." Statement of Financial Accounting Standards No. 69. Stamford, Conn., November 1982.

Financial Accounting Standards Board. "Effect of Rate Regulation on Accounting for Regulated Enterprises." FASB Discussion Memorandum. Stamford, Conn., December 31, 1979.

Financial Accounting Standards Board. "Extinguishment of Debt and the Offsetting of Restricted Assets Against Related Debt." Proposed Statement of Financial Accounting Standards. Stamford, Conn., October 13, 1982.

Financial Accounting Standards Board. "Field Tests of Financial Reporting in Units of General Purchasing Power." Research Report. Stamford, Conn., May 1977.

Financial Accounting Standards Board. "Financial Accounting and Reporting by Oil and Gas Producing Companies." Proposed Statement of Financial Accounting Standards. Stamford, Conn., July 15, 1977.

Financial Accounting Standards Board. "Financial Accounting and Reporting by Oil and Gas Producing Activities." Statement of Financial Accounting Standards No. 19. Stamford, Conn., December 1977.

Financial Accounting Standards Board. "Financial Reporting and Changing Prices: Specialized Assets--Mining and Oil and Gas." Proposed Statement of Financial Accounting Standards. Stamford, Conn., April 21, 1980.

Financial Accounting Standards Board. "Financial Reporting and Changing Prices: Specialized Assets--Mining and Oil and Gas." Statement of Financial Accounting Standards No. 39. Stamford, Conn., October 1980.

Financial Accounting Standards Board. "Objectives of Financial Reporting by Business Enterprises." Statement of Financial Accounting Concepts No. 1. Stamford, Conn., 1978.

Financial Accounting Standards Board. Oil and Gas Task Group. Interim Report on FASB Exposure Drafts: Financial Reporting and Changing Prices and Constant Dollar Accounting. Stamford, Conn.: June 13, 1979.

Financial Accounting Standards Board. Project: Accounting by Oil and Gas Producing Companies for the Tax Benefits of Percentage Depletion. Stamford, Conn., 1982.

Financial Accounting Standards Board. "Suspension of Certain Accounting Requirements for Oil and Gas Producing Companies." Proposed Statement of Financial Accounting Standards. Stamford, Conn., November 7, 1978.

Financial Accounting Standards Board. "Suspension of Certain Accounting Requirements for Oil and Gas Producing Companies." Statement of Financial Accounting Standards No. 25. Stamford, Conn., February 1979.

Norton, John C., and Rowe, Donald A. Accounting and Auditing Guide for United Kingdom Oil and Gas Exploration and Production. London: Institute of Chartered Accountants in England and Wales, 1978.

OIL AND GAS
GOVERNMENT DOCUMENTS
1973-1982

"Financial Accounting and Reporting for Oil and Gas Producing Activities Pursuant to the Federal Securities Law and the Energy Policy and Conservation Act of 1975." 17 Code of Federal Regulations, Part 210.4-10.

U.S. Federal Power Commission. Statistics of Interstate Natural Gas Pipeline Companies. Washington: Government Printing Office, 1977.

U.S. Federal Power Commission. Uniform Systems of Accounts Prescribed for Natural Gas Companies (Classes A, B, C, and D) in Effect on May 29, 1974. Washington: Government Printing Office, 1974, 1975.

U.S. Federal Power Commission. Uniform System of Accounts Prescribed for Public Utilities and Licensees (Classes A, B, C, and D) Subject to the Provisions of the Federal Power Act, in Effect on April 1, 1973. Washington: Government Printing Office, 1977.

U.S. Securities and Exchange Commission. Accounting Changes by Oil and Gas Producers. Accounting Series Release No. 261. Washington: Government Printing Office, 1979.

U.S. Securities and Exchange Commission. Accounting Changes by Oil and Gas Producers. Accounting Series Release No. 300. Washington: Government Printing Office, February 26, 1981.

U.S. Securities and Exchange Commission. Adoption of Requirements for Financial Accounting and Reporting Practices for Oil and Gas Producing Activities. Accounting Series Release No. 253. Washington: Government Printing Office, August 31, 1978.

U.S. Securities and Exchange Commission. Disclosure of Oil and Gas Reserves and Operations--Amendments to Regulation S-K. Release 33-6008, December 19, 1978.

U.S. Securities and Exchange Commission. Disclosure of Oil and Gas Reserves and Operations--Proposed Amendments to Regulation S-K. Release 33-5967, August 31, 1978.

U.S. Securities and Exchange Commission. FC (Full-Cost) vs. SE (Successful-Efforts): A Study of a Proposed Accounting Change's Competitive Impact. Washington: Government Printing Office, February 20, 1978.

U.S. Securities and Exchange Commission. Financial Reporting by Oil and Gas Producers. Accounting Series Release No. 289. Washington: Government Printing Office, February 26, 1981.

U.S. Securities and Exchange Commission. Interpretation of Accounting Series Release No. 257. Washington: Government Printing Office, 1979.

U.S. Securities and Exchange Commission. Interpretations of Accounting Series Releases Nos. 253 (as Amended by ASR No. 257) and ASR No. 269--Financial Accounting and Disclosure Requirements for Oil and Gas Producers. Staff Accounting Bulletin No. 35. Washington: Government Printing Office, 1979.

U.S. Securities and Exchange Commission. Notice of Adoption of Amendments to Regulation S-X Requiring Disclosure of Certain Replacement Cost Data. Accounting Series Release No. 190. Washington: Government Printing Office, 1976.

U.S. Securities and Exchange Commission. Oil and Gas Producers--Disclosure Requirements. Release 33-6079, June 27, 1979.

U.S. Securities and Exchange Commission. Oil and Gas Producers--Full Cost Accounting Practices. Accounting Series Release No. 258. Washington: Government Printing Office, December 19, 1978.

U.S. Securities and Exchange Commission. Oil and Gas Producers--Full Cost Accounting Practices. Release 33-6029, February 23, 1979.

U.S. Securities and Exchange Commission. Oil and Gas Producers--Postponement of Audit Requirements for Reserve Information. Accounting Series Release No. 270. Washington: Government Printing Office, September 1979.

U.S. Securities and Exchange Commission. Oil and Gas Producers--Proposed Supplemental Earnings Summary. Release 33-6080, June 27, 1979.

U.S. Securities and Exchange Commission. Oil and Gas Producers--Supplemental Disclosures on the Basis of Reserve Recognition Accounting. Accounting Series Release No. 269. Washington: Government Printing Office, September 1979.

U.S. Securities and Exchange Commission. Oil and Gas Reserve Information--Postponement of Audit Requirement. Accounting Series Release No. 277. Washington: Government Printing Office, April 17, 1980.

U.S. Securities and Exchange Commission. Proposal--Oil and Gas Producers Full Cost Accounting Practices. Release 33-5968, August 31, 1978.

U.S. Securities and Exchange Commission. Proposal--Supplemental Earnings Summary for Oil and Gas Producing Activities. Release 33-5969, August 31, 1978.

U.S. Securities and Exchange Commission. Requirements for Financial Accounting and Reporting Practices for Oil and Gas Producing Activities. Accounting Series Release No. 257. Washington: Government Printing Office, December 19, 1978.

U.S. Securities and Exchange Commission. Staff Accounting Bulletin No. 36. Washington: Government Printing Office, December 1979.

U.S. Securities and Exchange Commission. Staff Accounting Bulletin No. 41. Washington: Government Printing Office, February 1981.

U.S. Securities and Exchange Commission. Staff Accounting Bulletin No. 47. Washington: Government Printing Office, September 16, 1982.

U.S. Securities and Exchange Commission. Supplemental Disclosures of Oil and Gas Producing Activities. Releases 33-6444, 34-19337, 35-22776, IC-12898, FR-9, S7-936. Washington: Government Printing Office, December 1982.

OIL AND GAS
COPAS BULLETINS AND INTERPRETATIONS
1973-1982

Council of Petroleum Accountants Societies. "Accounting Procedure--Artic Operations." COPAS Bulletin No. 14. Tulsa: Kraftbilt, 1975.

Council of Petroleum Accountants Societies. "Accounting Procedure--Joint Operations." COPAS Bulletin No. 13. Tulsa: Kraftbilt, 1975.

Council of Petroleum Accountants Societies. "Accounting Procedure--Offshore Joint Operations." COPAS Bulletin No. 15. Tulsa: Kraftbilt, 1977.

Council of Petroleum Accountants Societies. "Coating and Wrapping Costs." COPAS Interpretation No. 7. Tulsa: Kraftbilt, April 1980.

Council of Petroleum Accountants Societies. "Computer Production Control Accounting Guidelines." COPAS Bulletin No. 12. Tulsa: Kraftbilt, 1975.

Council of Petroleum Accountants Societies. "COPAS Forms." COPAS Bulletin No. 4, revised. Tulsa: Kraftbilt, 1974.

Council of Petroleum Accountants Societies. COPAS Refinery and Marketing Exchange Accounting Bulletin No. 1. Tulsa: Kraftbilt, 1978.

Council of Petroleum Accountants Societies. "Definition of Railway Receiving Point." COPAS Interpretation No. 2. Tulsa: Kraftbilt, April 1980.

Council of Petroleum Accountants Societies. "Distribution of Boat and Fuel Expenses--Offshore Operations." COPAS Bulletin No. 18. Tulsa: Kraftbilt, May 1981.

Council of Petroleum Accountants Societies. "Employee Benefits Includable Within Operator's Percentage Calculation Charged to Joint Operations." COPAS Interpretation No. 11. Tulsa: Kraftbilt, October 1981.

Council of Petroleum Accountants Societies. "Employee Benefits Limitation." COPAS Interpretation No. 12. Tulsa: Kraftbilt, October 1982.

Council of Petroleum Accountants Societies. "Equalization of Truckers' Invoices on Transferred Material." COPAS Interpretation No. 6. Tulsa: Kraftbilt, April 1980.

Council of Petroleum Accountants Societies. "Gas Accounting

Manual." COPAS Bulletin No. 7, revised. Tulsa: Kraftbilt, January 1981.

Council of Petroleum Accountants Societies. "Joint Interest Audits in the Petroleum Industry: Guides to Protocol and Procedures." COPAS Bulletin No. 3, revised. Tulsa: Kraftbilt, October 1980.

Council of Petroleum Accountants Societies. "Loading, Transportation, and Unloading Costs. COPAS Interpretation No. 6, revised. Tulsa: Kraftbilt, June 1982.

Council of Petroleum Accountants Societies. "Material Classification." COPAS Bulletin No. 6, revised. Tulsa: Kraftbilt, 1982.

Council of Petroleum Accountants Societies. "Oil Accounting Manual." COPAS Bulletin No. 17. Tulsa: Kraftbilt, January 1981.

Council of Petroleum Accountants Societies. "Overhead--Joint Operations." COPAS Bulletin No. 16. Tulsa: Kraftbilt, October 1980.

Council of Petroleum Accountants Societies. "Premium Priced Material." COPAS Interpretation No. 10. Tulsa: Kraftbilt, May 1981.

Council of Petroleum Accountants Societies. "Pricing of Casing, Tubing, and Drill Pipe Used for Purpose Other Than Originally Intended." COPAS Interpretation No. 5. Tulsa: Kraftbilt, April 1980.

Council of Petroleum Accountants Societies. "Pricing of Line Pipe Movements Less Than 30,000 Pounds." COPAS Interpretation No. 4. Tulsa: Kraftbilt, April 1980.

Council of Petroleum Accountants Societies. "Pricing Tubular Goods at Eastern Mill Base Price vs. Lowest Prevailing Price." COPAS Interpretation No. 1. Tulsa: Kraftbilt, April 1980.

Council of Petroleum Accountants Societies. "Repricing of Transferred Mill Rejects and Limited Service Tubular Goods." COPAS Interpretation No. 8. Tulsa: Kraftbilt, April 1980.

Council of Petroleum Accountants Societies. "Selection of Mode of Hauling for Computing Freight--Mill to Railway Receiving Point." COPAS Interpretation No. 3. Tulsa: Kraftbilt, April 1980.

Council of Petroleum Accountants Societies. "Zone Priced Material." COPAS Interpretation No. 9. Tulsa: Kraftbilt, April 1980.

OIL AND GAS PROCEEDINGS 1973-1982

GENERAL

Adkerson, Richard C. "SEC and Financial Reporting by Oil and Gas Producing Companies," in Institute on Oil and Gas Accounting, 14th, Southwestern Legal Foundation, 1978 Proceedings. New York: Matthew Bender & Company, 1979, 1-1--1-87.

Ashabranner, G. D. "The Oil and Gas Royalty Clause--One-Eighth of What?" in Rocky Mountain Mineral Law Institute, 20th, 1974, Proceedings. New York: Matthew Bender & Company, 1975, 163-196.

Barton, M. Frank; Calhoun, Charles H.; and Carper, William Brent. "Oil Analysts Speak Out on Financial Accounting and Reporting in the Oil and Gas Industries: An Empirical Study," in Proceedings of the 32nd Annual Meeting, American Accounting Association, Southeast Regional Group. Sarasota, Fla., 1980, 47-50.

Beckler, Richard W. "Corporate Governance, the Foreign Corrupt Practices Act, and the SEC," in Institute on Oil and Gas Accounting, 15th, Southwestern Legal Foundation, 1979 Proceedings. New York: Matthew Bender & Company, 1980, 9-1--9-11.

Beta Alpha Psi. Current Issues in Accounting--Student Papers, Presented at Regional Meetings 1978-1979. Doris M. Cook, ed. Sarasota, Fla: National Council of Beta Alpha Psi, 1979.

Brown, Victor H. "Accounting for Inflation--How Relevant is Replacement Cost in Measuring the Impact of Inflation on the Oil Industry," in Accounting Papers, 31st Conference of Accountants. University of Tulsa, 1977, 5-17.

Brown, Victor H. "Replacement Cost Accounting," in Institute on Oil and Gas Accounting, 13th, Southwestern Legal Foundation, 1977 Proceedings. New York: Matthew Bender & Company, 1978, 3-1--3-18.

Castleberry, Jack C. "Impact of Government Regulations and Reporting Requirements on Petroleum Industry Accounting." in Accounting Papers, 29th Conference of Accountants. University of Tulsa, 1975, 33-42.

Deakin, Edward B., and Welsch, Glenn A. "Financial Reporting in the Oil and Gas Industry," in Oil and Gas Accounting Conference, University of New Orleans, 1977.

Field, Robert E. "Comment on Comparability of Divergent Financial Statements in the Petroleum Exploration Industry by S. Sunder." in Ross Institute of Accoúnting Research Proceedings on Topical Research in Accounting. New York, 1976, 266-272.

Hall, James D. "Recent Developments Regarding Financial Disclosure and Accounting Policies from the SEC," in Institute on Oil and Gas Accounting, 16th, Southwestern Legal Foundation, 1980 Proceedings. New York: Matthew Bender & Company, 1981, 9-1--9-11.

Kirk, Donald J. "Relationship Among Economic Realities, Economic Consequences, Accounting Conventions, and Accounting Standards-Setting Process," in Institute on Oil and Gas Accounting, 14th, Southwestern Legal Foundation, 1978 Proceedings. New York: Matthew Bender & Company, 1979, 8-1--8-15.

Klingstedt, John P. "Value Accounting in the Petroleum Industry," in Collected Papers of the Annual Meeting, American Accounting Association, 1975. Sarasota, Fla., 452-462.

Koester, Robert J., and Wilson, James L. "Recent Developments in Financial Accounting and Reporting by Oil and Gas Producing Companies," in Collected Papers of the Annual Meeting, American Accounting Association, 1978. Sarasota, Fla., 271-283.

Levy, Louis. "FASB Developments--Late and Forthcoming Pronouncements," in Institute on Securities Law and Regulation, 1st, New York University, 1977 Proceedings. New York, 1977, 225-238.

March, John W. "The FASB in 1979--An Overview," in Institute on Oil and Gas Accounting, 15th, Southwestern Legal Foundation, 1979 Proceedings. New York: Matthew Bender & Company, 1980, 8-1--8-16.

McDonald, Randal B. "FASB Exposure Draft on Financial Accounting and Reporting in the Extractive Industries," in Institute on Oil and Gas Accounting, 13th, Southwestern Legal Foundation, 1977 Proceedings. New York: Matthew Bender & Company, 1978, 1-1--1-25.

Moore, C. H. "Conflict of Petroleum Accounting Practices and New Relationship with Lawmakers," in Accounting Papers, 30th Conference of Accountants. University of Tulsa, 1976, 21-36.

Norr, David. "Comment on Comparability of Divergent Financial Statements in the Petroleum Exploration Industry by S. Sunder." in Ross Institute of Accounting Proceedings on Topical Research in Accounting. New York, 1976, 273-284.

Nussbaum, James G., and Huffman, James T. "Accounting and Reporting Practices in the Petroleum Industry," in Rocky Mountain Mineral Law Institute, 23rd, 1977 Proceedings. New York: Matthew Bender & Company, 1977, 693-735.

Pacter, Paul A. "Will the FASB Survive the Current Pressures?" in Accounting Papers, 32nd Conference of Accountants, University of Tulsa, 1978, 66-84.

Petrie, Thomas A. "Public Energy Companies' Disclosure of Financial and Related Information--How a Financial Analyst Views the Data Available," in Institute on Oil and Gas Accounting, 16th, Southwestern Legal Foundation, 1980 Proceedings. New York: Matthew Bender & Company, 6-1--6-8.

Smith, Philip S. "Oil and Gas: Comment on Interim Task Group Report," in Conference on Financial Reporting and Changing Prices, 1979 Proceedings. Stamford, Conn., 1979, 93-97.

Sunder, Shyam. "Comparability of Divergent Financial Statements in the Petroleum Exploration Industry." in Ross Institute of Accounting Research Proceedings on Topical Research in Accounting. New York, 1976, 251-265.

Thompson, Robert C. "Public Energy Companies' Disclosure of Financial and Related Information--What a Company's Management Feels Is Important," Institute on Oil and Gas Accounting, 16th, Southwestern Legal Foundation, 1980 Proceedings. New York: Matthew Bender & Company, 1981, 5-1--5-11.

Vangermeersch, Richard. "Reliving the Oil and Gas Accounting Mistakes of the Past," in Collected Papers of the Annual Meeting, American Accounting Association, 1979. Sarasota, Fla., 1980, 708-717.

DISCLOSURES AND RESERVE RECOGNITION ACCOUNTING

Adkerson, Richard C. "Reserve Recognition Accounting Revisited: A Review of 1979 Oil and Gas Disclosures by Public Companies," in The Economic and Evaluation Symposium, February 1981.

Alexander, Michael O.; Bloyd, Dean M.; Moore, C. H.; Sampson, A. Clarence; and Smith, Philip S. "Disclosures in Financial Statements of Oil and Gas Companies--A Panel Discussion," in Institute on Oil and Gas Accounting, 17th, Southwestern Legal Foundation, 1981 Proceedings. New York: Matthew Bender & Company, 1982, 6-1--6-57.

Barton, Larry M. "The Auditor's Examination of Oil and Gas Reserve Information," in Western Region Convention, American Accounting Association. Sarasota, Fla., 1979.

Deakin, Edward B. "Reporting Current Cost for Oil and Gas Reserves--Part II," in Institute on Oil and Gas Accounting, 13th, Southwestern Legal Foundation, 1977 Proceedings. New York: Matthew Bender & Company, 1978, 2-7--2-14.

Deakin, E. B. "Reserve Recognition Accounting: How It Came About and Where It Is Headed," in Institute on Oil and Gas Accounting, 16th, Southwestern Legal Foundation, 1980 Proceedings. New York: Matthew Bender & Company, 1981, 1-1--1-12.

Dickman, J. Jerry. "SEC Requirement of Disclosure," in Institute on Oil and Gas Law and Taxation, 26th, Southwestern Legal Foundation, 1975 Proceedings. New York: Matthew Bender & Company, 1975, 239-265.

Garrett, Ray, Jr. "The SEC Involvement in the Disclosure of Energy Reserves," in Institute on Oil and Gas Law and Taxation, 27th, Southwestern Legal Foundation, 1976 Proceedings. New York: Matthew Bender & Company, 1976, 97-118.

Gilmore, Robert B. "Public Reporting of Oil and Gas Reserves--Public Companies' Reports of Petroleum Reserves and Values Can Further Confuse Users of Energy Data (Part 2)," in Institute on Oil and Gas Accounting, 15th, Southwestern Legal Foundation, 1979 Proceedings. New York: Matthew Bender & Company, 1980, 4-1--4-10.

Hall, James D. "Recent Developments Regarding Financial Disclosure and Accounting Policies from the SEC," in Institute on Oil and Gas Accounting, 16th, Southwestern Legal Foundation, 1980 Proceedings. New York: Matthew Bender & Company, 1981, 9-1--9-11.

Halverson, James T. "An Analysis of the Oil and Natural Gas Reserve Reporting Problems: The Government's Need to Know Versus the Private Company's Need to Protect the Confidentiality of Its Sensitive Business Information," in Institute on Oil and Gas Law and Taxation, 27th, Southwestern Legal Foundation, 1976 Proceedings. New York: Matthew Bender & Company, 1976, 119-136.

Huddleston, B. P. "Public Statements of Fair Market Value of Oil and Gas Reserves," in Institute on Oil and Gas Accounting, 16th, Southwestern Legal Foundation, 1980 Proceedings. New York: Matthew Bender & Company, 1981, 2-1--2-12.

Klingstedt, John P. "Value Accounting in the Petroleum Industry," in Collected Papers of the Annual Meeting, American Accounting Association, 1975. Sarasota, Fla., 452-462.

LeGrange, U. J. "Reserve Recognition Accounting--An Experiment or a Foregone Conclusion? (Part 2)," in Institute on Oil and Gas Accounting, 15th, Southwestern Legal Foundation, 1979 Proceedings. New York: Matthew Bender & Company, 1980, 2-1--2-7.

May, Alan, Jr. "Public Reporting of Oil and Gas Reserves--Auditing Oil and Gas Reserve Information (Part 1)," in Institute on Oil and Gas Accounting, 15th, Southwestern Legal Foundation, 1979 Proceedings. New York: Matthew Bender & Company, 1980, 3-1--3-11.

Petrie, Thomas A. "Public Energy Companies' Disclosure of Financial and Related Information--How a Financial Analyst Views the Data Available," in Institute on Oil and Gas Accounting, 16th, Southwestern Legal Foundation, 1980 Proceedings. New York: Matthew Bender & Company, 6-1--6-8.

Sampson, A. Clarence. "Reserve Recognition Accounting--Current Status of Reserve Recognition Accounting (Part 1)," in Institute on Oil and Gas Accounting, 15th, Southwestern Legal Foundation, 1979 Proceedings. New York: Matthew Bender & Company, 1980, 1-1--1-10.

Thompson, Robert C. "Public Energy Companies' Disclosure of Financial and Related Information--What a Company's Management Feels Is Important," Institute on Oil and Gas Accounting, 16th, Southwestern Legal Foundation, 1980 Proceedings. New York: Matthew Bender & Company, 1981, 5-1--5-11.

Welsch, Glenn A. "Reporting Current Cost for Oil and Gas Reserves--Part I," in Institute on Oil and Gas Accounting, 13th, Southwestern Legal Foundation, 1977 Proceedings. New York: Matthew Bender & Company, 1978, 2-1--2-6.

DRILLING FUNDS, PROGRAMS, JOINT VENTURES, AND PARTNERSHIPS

Crichton, Thomas, IV, and Griffin, Campbell A., Jr. "Securities Problems and Tax Implications of Oil and Gas Investments by Non-Industry Financiers," in Rocky Mountain Mineral Law Institute, 27th, 1981 Proceedings. New York: Matthew Bender & Company, 1982, 1333-1406.

Dutton, Granville. "Accounting Procedures: Contracts or Controversies?" in Rocky Mountain Mineral Law Institute, 19th, 1973 Proceedings. New York: Matthew Bender & Company, 1974, 117-154.

Hill, Rodney C. "Oil Programs--Capital Source for the Oil Industry," in Exploration and Economics of the Petroleum Industry, 11th, Southwestern Legal Foundation, 1973 Proceedings. New York: Matthew Bender & Company, 1973, 155-169.

McCollam, John M. "A Selective Comparison of Contractual Operating Problems Under Federal Offshore and Onshore Oil and Gas Leases," in Institute on Oil and Gas Law and Taxation, 29th, Southwestern Legal Foundation, 1978 Proceedings. New York: Matthew Bender & Company, 1978, 229-331.

Twenhafel, Marvin G. "Oil-Gas Division Orders: Their Origin, Varieties, and Usage," in Rocky Mountain Mineral Law Institute, 27th, 1981 Proceedings. New York: Matthew Bender & Company, 1982, 1479-1515.

Wigley, Marvin L. "AAPL Form 610-1977 Model Form Operating Agreement," in Rocky Mountain Mineral Law Institute, 24th, 1978 Proceedings. New York: Matthew Bender & Company, 1978, 693-716.

Young, J. O. "Oil and Gas Operating Agreements: Producers 88 Operating Agreements, Selected Problems and Suggested Solutions," in Rocky Mountain Mineral Law Institute, 20th, 1974 Proceedings. New York: Matthew Bender & Company, 1975, 197-225.

Young, J. O. "Oil and Gas Operations: Who Does What, to Whom, for Whom and Who Pays, How, and When," in Rocky Mountain Mineral Law Institute, 27th, 1981 Proceedings. New York: Matthew Bender & Company, 1982, 1651-1691.

ECONOMICS, FINANCE, AND EVALUATION

Brennan, Richard S. "Current Trends in Oil and Gas Financing," in Rocky Mountain Mineral Law Institute, 25th, 1979 Proceedings. New York: Matthew Bender & Company, 1979, 16-1--16-20.

Burris, Anthony E., and Robson, William M. "Evaluation of Royalty Bid and Profit Share Bid Bidding Systems," in Exploration and Economics of the Petroleum Industry, 20th, Southwestern Legal Foundation, 1982 Proceedings. New York: Matthew Bender & Company, 1982, 123-172.

Cozzolins, John, M. "Exploration Risk Management," in Exploration and Economics of the Petroleum Industry, 19th, Southwestern Legal Foundation, 1981 Proceedings. New York: Matthew Bender & Company, 1981, 97-116.

Crichton, Thomas, IV, and Griffin, Campbell A., Jr. "Securities Problems and Tax Implications of Oil and Gas Investments by Non-Industry Financiers," in Rocky Mountain Mineral Law Institute, 27th, 1981 Proceedings. New York: Matthew Bender & Company, 1982, 1333-1406.

Davis, W. Kenneth. "New Directions in United States Energy Policy," in Institute on Oil and Gas Accounting, 17th, Southwestern Legal Foundation, 1981 Proceedings. New York: Matthew Bender & Company, 1982, 1-1--1-7.

Hill, Rodney C. "Oil Programs--Capital Source for the Oil Industry," in Exploration and Economics of the Petroleum Industry, 11th, Southwestern Legal Foundation, 1973 Proceedings. New York: Matthew Bender & Company, 1973, 155-169.

Hunt, John F., and Camp, Herbert L. "Project Financing--Oil and Gas Ventures," in Institute on Oil and Gas Law and Taxation, 27th, Southwestern Legal Foundation, 1976 Proceedings. New York: Matthew Bender & Company, 1976, 215-258.

Larue, John D. "The Rising Cost of New U.S. Oil Reserves, 1959-1975," in Exploration and Economics of the Petroleum Industry, 14th, Southwestern Legal Foundation, 1976 Proceedings. New York: Matthew Bender & Company, 1976, 193-238.

Matthews, T. K. "Project Financing Modes for the North Sea and Applications to Other Areas," in Exploration and Economics of the Petroleum Industry, 14th, Southwestern Legal Foundation, 1976 Proceedings. New York: Matthew Bender & Company, 1976, 239-250.

Monteith, Edward E., Jr. "Project Financing--New Uses and New Risks," in Institute on Oil and Gas Accounting, 13th, Southwestern Legal Foundation, 1977 Proceedings. New York: Matthew Bender & Company, 1978, 4-1--4-12.

Pickens, T. Boone, Jr. "Optimum Reserve Base for Petroleum Companies," in Exploration and Economics of the Petroleum Industry, 20th, Southwestern Legal Foundation, 1982 Proceedings. New York: Matthew Bender & Company, 1982, 267-271.

Swearingen, John E. "Profits: The Key to the Energy Storehouse," in Exploration and Economics of the Petroleum Industry, 12th, Southwestern Legal Foundation, 1974 Proceedings. New York: Matthew Bender & Company, 1974, 45-59.

PLANNING, CONTROL, AND BUDGETING

Cozzolins, John, M. "Exploration Risk Management," in Exploration and Economics of the Petroleum Industry, 19th, Southwestern Legal Foundation, 1981 Proceedings. New York: Matthew Bender & Company, 1981, 97-116.

Lawrence, G. G. "Answering Audit Recommendations to Engineer Work Standards," in American Institute of Industrial Engineers Proceedings, 1982 Conference, 671-678.

Pannill, William. "Strategies for Maximizing Revenues Under the Natural Gas Policy Act of 1978," in Institute on Oil and Gas Accounting, 15th, Southwestern Legal Foundation, 1979 Proceedings. New York: Matthew Bender & Company, 1980, 7-1--7-12.

Pickens, T. Boone, Jr. "Optimum Reserve Base for Petroleum Companies," in Exploration and Economics of the Petroleum Industry, 20th, Southwestern Legal Foundation, 1982 Proceedings. New York: Matthew Bender & Company, 1982, 267-271.

INTERNATIONAL

Matthews, T. K. "Project Financing Modes for the North Sea and Applications to Other Areas," in Exploration and Economics of the Petroleum Industry, 14th, Southwestern Legal Foundation, 1976 Proceedings. New York: Matthew Bender & Company, 1976, 239-250.

Most, Kenneth S. "Financial Reporting by the Oil and Gas Industry in Europe," in Proceedings of the 32nd Annual Meeting, American Accounting Association, Southeast Regional Group, Sarasota, Fla., 1980, 148-153.

AUDITING

Barton, Larry M. "The Auditor's Examination of Oil and Gas Reserve Information," in Western Region Convention, American Accounting Association. Sarasota, Fla., 1979.

Beck, David J. "Department of Energy Audit Issues: Crude Oil Issues," in Institute on Oil and Gas Law and Taxation, 30th, Southwestern Legal Foundation, 1979 Proceedings. New York: Matthew Bender & Company, 1979, 1-42.

Lawrence, G. G. "Answering Audit Recommendations to Engineer Work Standards," in American Institute of Industrial Engineers Proceedings, 1982 Conference, 671-678.

May, Alan, Jr. "Public Reporting of Oil and Gas Reserves--Auditing Oil and Gas Reserve Information (Part 1)," in Institute on Oil and Gas Accounting, 15th, Southwestern Legal Foundation, 1979 Proceedings. New York: Matthew Bender & Company, 1980, 3-1--3-11.

Montgomery, Robert E., Jr. "Department of Energy Audit Issues: Recent Developments in DOE Audits of Refiners and Marketers," in Institute on Oil and Gas Law and Taxation, 30th, Southwestern Legal Foundation, 1979 Proceedings. New York: Matthew Bender & Company, 1979, 61-96.

Morse, Richard C. "Department of Energy Audit Issues: Natural Gas Liquids: Selected Problems in Regulation," in Institute on Oil and Gas Law and Taxation, 30th, Southwestern Legal Foundation, 1979 Proceedings. New York: Matthew Bender & Company, 1979, 43-59.

Williams, Richard T. "Strategies for Responding to ERA Audits and Enforcement Proceedings," in Institute on Oil and Gas Accounting, 14th, Southwestern Legal Foundation, 1978 Proceedings. New York: Matthew Bender & Company, 1979, 5-1--5-12.

GOVERNMENT REGULATION AND PRICE CONTROL

Allott, Gordon L., Jr. "Crude Oil Deregulation," in Rocky Mountain Mineral Law Institute, 25th, 1979 Proceedings. New York: Matthew Bender & Company, 1979, 15-1--15-15.

Andersen, Arthur T. "Energy Information: A Perspective on the Energy Information Administration and the Financial Reporting System," in Institute on Oil and Gas Accounting, 16th, Southwestern Legal Foundation, 1980 Proceedings. New York: Matthew Bender & Company, 1981, 3-1--3-19.

Bachman, Kenneth L., Jr. "Developments and Trends in DOE Enforcement," in Institute on Oil and Gas Law and Taxation, 31st, Southwestern Legal Foundation, 1980 Proceedings. New York: Matthew Bender & Company, 1980, 61-85.

Barnes, Gerald H. "Update on DOE's Oil Pricing Regulations," in Institute on Oil and Gas Law and Taxation, 32nd, Southwestern Legal Foundation, 1981 Proceedings. New York: Matthew Bender & Company, 1981, 267-299.

Beck, David J. "Department of Energy Audit Issues: Crude Oil Issues," in Institute on Oil and Gas Law and Taxation, 30th, Southwestern Legal Foundation, 1979 Proceedings. New York: Matthew Bender & Company, 1979, 1-42.

Castleberry, Jack C. "Impact of Government Regulations and Reporting Requirements on Petroleum Industry Accounting." in Accounting Papers, 29th Conference of Accountants. University of Tulsa, 1975, 33-42.

Craven, Donald B. "New Dimensions in Federal Regulation of Crude Oil and Petroleum Products Under the Department of Energy," in Institute on Oil and Gas Law and Taxation, 29th, Southwestern Legal Foundation, 1978 Proceedings. New York: Matthew Bender & Company, 1978, 1-37.

Fuller, Keith L. "Profit Maximization Under DOE Crude Oil Regulations," in Institute on Oil and Gas Accounting, 14th, Southwestern Legal Foundation, 1978 Proceedings. New York: Matthew Bender & Company, 1979, 6-1--6-32.

Henke, Michael J. "Natural Gas Liquid Pricing Under DOE Regulations," in Rocky Mountain Mineral Law Institute, 26th, 1980 Proceedings. New York: Matthew Bender & Company, 1980, 857-886.

Henke, Michael J., and Anderson, Eric G. "Current Issues Concerning DOE Regulation of Natural Gas Liquids," in Institute on Oil and Gas Law and Taxation, 31st, Southwestern Legal Foundation, 1980 Proceedings. New York: Matthew Bender & Company, 1980, 27-59.

Lang, Scott H. "DOE Organization: The Limits of Regulatory Power," in Institute on Oil and Gas Law and Taxation, 30th, Southwestern Legal Foundation, 1979 Proceedings. New York: Matthew Bender & Company, 1979, 119-153.

Langdon, Jim C., Jr. "Domestic Crude Oil Production--The FEA Regulatory Framework," in Institute on Oil and Gas Law and Taxation, 28th, Southwestern Legal Foundation, 1977 Proceedings. New York: Matthew Bender & Company, 1977, 1-63.

Langdon, Jim C., Jr. "FEA Price Controls for Crude Oil and Refined Petroleum Products," in Institute on Oil and Gas Law and Taxation, 26th, Southwestern Legal Foundation, 1975 Proceedings. New York: Matthew Bender & Company, 1975, 55-100.

Langdon, Jim. "Strategies for Maximizing Revenues Under DOE Oil Price Regulations," in Institute on Oil and Gas Accounting, 15th, Southwestern Legal Foundation, 1979 Proceedings. New York: Matthew Bender & Company, 1980, 6-1--6-16.

Lavenant, Rene P., Jr. "Operator's Liability for Overcharges in the Sale of Production," in Institute on Oil and Gas Law and Taxation, 31st, Southwestern Legal Foundation, 1980 Proceedings. New York: Matthew Bender & Company, 1980, 131-176.

Ligon, Duke R. "Crude Oil Pricing: Current Regulations and the Shift to Decontrol," in Institute on Oil and Gas Law Taxation, 31st, Southwestern Legal Foundation, 1980 Proceedings. New York: Matthew Bender & Company, 1980, 1-25.

Ligon, Duke R. "The Federal Energy Administration: Significant Developments in Price and Allocation Regulation; Expanding Horizons in Energy Planning," in Institute on Oil and Gas Law and Taxation, 27th, Southwestern Legal Foundation, 1976 Proceedings. New York: Matthew Bender & Company, 1976, 137-175.

Methvin, Gaynell C. "Department of Energy v. Oil Industry--Recent Developments in Litigation Over Pricing Regulations," in Institute on Oil and Gas Accounting, 16th, Southwestern Legal Foundation, 1980 Proceedings. New York: Matthew Bender & Company, 1981, 8-1--8-16.

Montgomery, Robert E., Jr. "Department of Energy Audit Issues: Recent Developments in DOE Audits of Refiners and Marketers," in Institute on Oil and Gas Law and Taxation, 30th, Southwestern Legal Foundation, 1979 Proceedings. New York: Matthew Bender & Company, 1979, 61-96.

Morse, Richard C. "Department of Energy Audit Issues: Natural Gas Liquids: Selected Problems in Regulation," in Institute on Oil and Gas Law and Taxation, 30th, Southwestern Legal Foundation, 1979 Proceedings. New York: Matthew Bender & Company, 1979, 43-59.

Peach, J. Dexter, "GAO's Role in Energy Policy," in Institute on Oil and Gas Accounting, 14th, Southwestern Legal Foundation, 1978 Proceedings. New York: Matthew Bender & Company, 1979, 2-1--2-11.

Poland, Sherman S. "Contrasting Administrative Procedures Before the Department of Energy," in Institute on Oil and Gas Law and Taxation, 30th, Southwestern Legal Foundation, 1979 Proceedings. New York: Matthew Bender & Company, 1979, 97-118.

Richardson, Thomas A. "Crude Oil Pricing--Current Regulations and Practices," in Rocky Mountain Mineral Law Institute, 23rd, 1977 Proceedings. New York: Matthew Bender & Company, 1977, 803- 824.

Wakefield, Stephen A. "Allocation, Price Control and the FEA: Regulatory Policy and Practice in the Political Arena," in Rocky Mountain Mineral Law Institute, 21st, 1975 Proceedings. New York: Matthew Bender & Company, 1975, 257-284.

Williams, Richard T. "Strategies for Responding to ERA Audits and Enforcement Proceedings," in Institute on Oil and Gas Accounting, 14th, Southwestern Legal Foundation, 1978 Proceedings. New York: Matthew Bender & Company, 1979, 5-1--5-12.

REFINING

Montgomery, Robert E., Jr. "Department of Energy Audit Issues: Recent Developments in DOE Audits of Refiners and Marketers," in Institute on Oil and Gas Law and Taxation, 30th, Southwestern Legal Foundation, 1979 Proceedings. New York: Matthew Bender & Company, 1979, 61-96.

PIPELINING AND TRANSMISSION

Upchurch, Claude E. "Split Stream Gas Sales and the Gas Storage and Balancing Agreement," in Rocky Mountain Mineral Law Institute, 24th, 1978 Proceedings. New York: Matthew Bender & Company, 1978, 665-692.

OIL AND GAS
HEARINGS, LECTURES, SPEECHES
1973-1982

Arthur Andersen & Company. Accounting for Oil and Gas Exploration and Development Costs. Before the Committee on Extractive Industries of the Accounting Principles Board of the American Institute of Certified Public Accountants, November 22-23, 1971. Chicago, 1971.

Financial Accounting Standards Board. Discussion Memorandum on Financial Accounting and Reporting in the Extractive Industries. Vol. 7 of transcript of public hearing. March 30-31, April 1, 4, 1977. Stamford, Conn., 1977.

Kirk, Donald J. Accounting Practices, Oil and Gas Producers. SEC Public Hearing, March 29, 1978. Stamford, Conn.: Financial Accounting Standards Board, 1978.

Nichols, Robert G. "Soft Data Risks and Rewards." CPA Journal 49 (December 1979): 5-6.

Phillips, John N.; Flore, Louis R.; and Weinig, R. W. Remarks to the New York Society of Security Analysts. June 5, 1975. Boston: Eastern Gas and Fuel Associates, 1975.

Porter, Stanley P. Successful Efforts Accounting for Extractive Industries. Before the Department of Energy, February 22, 1978. New York: Arthur Young & Company, 1978.

United States. Securities and Exchange Commission. Opening Statement of the Chairman--Public Hearing in the Matter of Accounting Practices for Oil and Gas Producers. March 29, 1978. Washington, 1978.

Way, Alva O. Testimony Before the Securities and Exchange Commission. March 29, 1978. Stamford, Conn.: Financial Accounting Foundation, 1978.

OIL AND GAS
NEWS RELEASES AND SHORT COMMENTS
1973-1982

GENERAL

"Apples and Oranges and Oil." Forbes 112 (December 1, 1973): 49-50.

"The Numbers Game--Will the Oil Companies Try to Hide Their Profits?" Forbes 113 (February 1, 1974): 40-41.

"A Bookkeeping Wage Behind Oil Profits." Business Week (May 4, 1974): 22-23.

"Texaco Inc. Will Disclose in Its 1974 Annual Report the Effect on Its Earnings of Certain Accounting Practices It Follows for Oil-Exploration Costs." Wall Street Journal, 14 (February 1975), 10.

"Financial Accounting Standards Board Proposing That Oil and Gas Companies Change Their Accounting to Show Larger Income-Tax Charges in Reports to Shareholders as a Result of Losing Percentage Depletion." Wall Street Journal, 1 May 1975, 15.

"Financial Accounting Standards Board Decided to Hold Public Hearing in September on Proposed Change in Oil and Gas Companies' Accounting to Allow for the Elimination of Percentage Depletion for Major Producers." Wall Street Journal, 27 June 1975, 19.

Beresford, Dennis R., and Groves, Ray J. "FASB Proposes That Deferred Taxes for Oil and Gas Intangible Drilling Cost Be Recorded." Financial Executive 43 (June 1975): 10.

"Financial Accounting Standards Board Set Sept. 10 for a Previously Announced Public Hearing on How Oil and Gas Producers Should Account for Income Taxes." Wall Street Journal, 16 July 1975, 7.

"A Rule That Could Cost Oilmen $3 Billion." Business Week (July 28, 1975): 76-77.

"FASB to Review Proposal for Accounting Treatment of Income Taxes by Oil and Gas Companies." SEC Accounting Report 1 (August 1975): 8.

Financial Accounting Standards Board Moved to Take Up Thorny Topic of Oil-and-Gas Accounting Before Congress Decides to Give the Job to Someone Else." Wall Street Journal, 6 October 1975, 15.

"Financial Accounting Standards Board Dropped Proposal That Many Oil and Gas Companies Contended Would Have Halved Their 1975 Earnings." Wall Street Journal, 16 October 1975, 17.

"FASB Adds Oil and Gas to Its Technical Agenda." Journal of Accountancy 140 (November 1975): 9.

"Accounting Principles for Extractive Industries on FASB Agenda." SEC Accounting Report 2 (December 1975): 8.

Hot Debate Over Oil Exploration Costs." Business Week (February 9, 1976): 81-83.

"The Hot Debate Over Oil Exploration Costs." Business Week (February 9, 1976): 81-83.

"FASB Appoints Task Force for Extractive Industries." Journal of Accountancy 141 (February 1976): 14, 18.

"FASB Tentatively Lumps Mining with Oil and Gas." Journal of Accountancy 141 (March 1976): 16.

"Now You See It." Forbes 118 (August 15, 1976): 38.

"The Numbers Game." Forbes 118 (August 15, 1976): 38.

"FASB Sets Hearing to Discuss Standards for Oil, Gas, Mining Industries." Wall Street Journal, December 30, 1976, 6.

Seidler, Lee J., and Gibson, Corine. "Oil and Gas Debate Will Live the New Year." Accounting Issues (Janaury 11, 1977): 17.

"Are They Earning It or Aren't They?" Forbes 119 (February 15, 1977): 88, 90.

"Discussion Memorandum Issued on Oil and Gas Accounting Rules." SEC Accounting Report 3 (February 1977): 7.

"FASB--Board Issues Memorandum on Extractive Industries." Journal of Accountancy 143 (February 1977): 7.

"FASB Begins Hearings on Hotly Debated Oil, Gas Industry Rules." Wall Street Journal, 30 March 1977, 4.

Symonds, Edward. "The Oil Industry Faces Accounting Shocks." Petroleum Economist (Eng.) 44 (May 1977): 181.

Norby, William C. "Oil and Gas Accounting." Financial Analysts Journal 33 (May/June 1977): 14, 16, 18, 29, 42.

Bisgay, Louis. "MAP Committee Comments on Extractive Industries Issues." Management Accounting 58 (June 1977): 67.

"Successful Efforts Method for Reporting Oil and Gas Search Costs Backed by FASB." Wall Street Journal, 20 July 1977, 8.

"Extractive Industries Discussion Memorandum." CPA Journal 47 (July 1977): 59-60.

"FASB: A Single Oil Standard." Business Week (August 1, 1977): 50.

"Oil and Gas Accounting: SEC Asks for Additional Comments." Journal of Accountancy 144 (August 1977): 22, 24, 28.

Symonds, Edward. "Agonizing Verdict." Accountant (Eng.) 177 (September 15, 1977): 316-317.

Beresford, Dennis R., and Neary, Robert D. "FASB Wants Uniform Oil and Gas Accounting." Financial Executive 45 (September 1977): 8, 10.

"Draft Calls for Successful Efforts Oil and Gas Accounting." Journal of Accountancy 144 (September 1977): 20, 24.

"An FASB Ruling Hurts Oil Exploration." Business Week (October 10, 1977): 41-42.

"Financial Accounting Standards Board Sees Threat to Its Authority on Oil, Gas Reports from Little-Noticed Amendment to Unrelated Bill Passed by Senate Oct. 6." Wall Street Journal, 17 October 1977, 10.

"Accounting Method Proposed for Oil and Gas Producers." CA Magazine (Canada) 110 (October 1977): 9-10.

Bisgay, Louis. "MAP Committee Files Committee with SEC." Management Accounting 59 (October 1977): 12, 64.

"Oil and Gas." CPA Journal 47 (October 1977): 86.

"Oil and Gas Accounting." CPA Journal 47 (October 1977): 86.

"Justice Dept. Asked SEC to Delay Adoption of a Uniform Accounting Method for Oil and Gas Producers Until SEC Has Assessed Competitive Effects of the Proposed Change." Wall Street Journal, 28 November 1977, 8.

Armstrong, Marshall S. "FASB, Moving Ahead, Asks Cooperation of the Press." Newspaper Controller 3 (November 1977): 7-8.

Beresford, Dennis R., and Neary, Robert D. "SEC Requires Oil and Gas Producers to Give Effects of FASB's Proposal." Financial Executive 45 (November 1977): 18, 20.

"Oil and Gas Accounting." CPA Journal 47 (November 1977): 88-89.

Norby, William C. "Oil and Gas Accounting Revisited." Financial Analysts Journal, 33 (November/December 1977): 18-20, 75.

"FASB Has Ruled That All of the Nation's Oil and Natural Gas Producers Must Use the So-Called 'Successful Efforts' Method of Accounting for the Costs of Searching for and Developing Oil and Gas Fields." Wall Street Journal, 6 December 1977, 19.

Symonds, Edward. "Replacing the Irreplaceable." Accountant (Eng.) 177 (December 15, 1977): 762-763.

Santry, David G. "Earnings Cut From Oil Exploration." Business Week (December 19, 1977): 67.

"SEC Ordered Public Hearings on a Controversial Proposal to Standarize the Accounting Methods That Oil and Gas Companies Use in Reports to the Government." Wall Street Journal, 29 December 1977, 3.

Beresford, Dennis R., and Neary, Robert D. "SEC Wants Oil and Gas Reserves to Be Valued; Proposes to Endorse Successful Efforts Accounting." Financial Executive 45 (December 1977): 12, 14.

Miller, Daniel E. "SEC." Financial Executive 45 (December 1977): 10.

"Will Congress Reject the FASB's Oil and Gas Accounting Methods?" SEC Accounting Report 4 (December 1977): 2-3.

"Storm Brewing Over Oil Accounts." Wall Street Journal, 22 February 1978, 22

Beresford, Dennis R., and Neary, Robert D. "Oil and Gas Producers Get FASB Rules; SEC to Consider Them." Financial Executive 46 (February 1978): 12.

Bisgay, Louis. "FASB Rules on Oil, Gas Accounting." Management Accounting 59 (February 1978): 10.

"FASB Issues Final Rules." SEC Accounting Report 4 (February 1978): 8.

Financial Accounting Standards Board. "Financial Accounting and Reporting by Oil and Gas Producing Companies." Statement of Financial Accounting Standards No. 19. Journal of Accountancy 145 (February 1978): 94-103.

"Oil and Gas Accounting: FASB Issues Final Rules." SEC Accounting Report 4 (February 1978): 8.

"SEC To Hold Hearings on FASB Rule on Oil and Gas Accounting." Journal of Accountancy 145 (February 1978): 20.

"Independents Oppose the FASB's Oil Rules." Business Week (March 6, 1978): 30-31.

"Battle Over Accounting Methods for Oil and Gas Producers Moved into a Critical Stage as the SEC Opened Hearings on the Controversy." Wall Street Journal, 30 March 1978, 7.

"Oil Accounts Should Be Uniform." Wall Street Journal, 31 March 1978, 14.

Norby, William C. "Oil and Gas Accounting." Financial Analysts Journal 34 (March/April 1978): 73.

Anreder, Steven S. "Those Bottom Lines." Barron's 58 (April 10, 1978): 4-5.

"Letters to the Editor Discussing Accounting Methods of Oil and Gas Accounts." Wall Street Journal, 24 April 1978, 19.

"AICPA Back FASB Statement of No. 19 in Letter to the SEC." Journal of Accountancy 145 (May 1978): 10.

"SEC Hearings." Journal of Accountancy 145 (May 1978): 10.

Symonds, Edward. "Pressure Mounts to Unify Oil Accounting Methods." Petroleum Economist (Eng.) 45 (May 1978): 185-189.

Kolins, Wayne. "Controversial Statement Issued for Oil and Accounting." Practical Accountant 11 (May/June 1978): 50-51.

"The FASB Has Made Public a Strong Rebuttal to Opponents of Its Controversial Ruling on Accounting for the Costs of Searching for and Developing Oil and Gas Fields." Wall Street Journal, 6 June 1978, 16.

Symonds, Edward. "Oil Companies at Loggerheads." Accountant (Eng.) 178 (June 8, 1978): 764-765.

Randall, Robert F. "SEC's Oil-Gas Hearings." Management Accounting 59 (June 1978): 9.

"Oil Together Now." Wall Street Journal, 18 July 1978, 20.

"FASB Counters Opponents of Its Oil and Gas Ruling." Journal of Accountancy 146 (July 1978): 8, 10.

"FASB Responses to Requests for Action." SEC Accounting Report 4 (July 1978): 7-8.

"A Puzzle for the SEC on Oil Accounting Rules." Business Week (August 7, 1978): 29.

"Bad Effects of FASB Statement 19 Denied." CA Magazine (Canada) 3 (August 1978): 14, 16.

"An Accounting Rule Takes Oil by Surprise." Business Week (September 11, 1978): 62.

Anreder, Steven S. "Changing the Ground Rules, Washington May Have Launched a Revolution in Accounting." Barron's 58 (September 25, 1978): 11-12, 32.

Owens, G. Chester. "Oil and Gas Rule Blowout." Journal of Accountancy 146 (September 1978): 104.

Symonds, Edward. "The Judgment of Harold." Accountant (Eng.) 179 (October 5, 1978): 425-426.

"The FASB Delayed Accounting Rule for Costs of Gas, Oil Exploration and Development." Wall Street Journal, 5 October 1978, 12.

Mims, Robert. "A Solution Fraught with Problems for Oil and Gas Accounting." Business Week (October 9, 1978) 105-106.

"Oil and Gas Accounting: SEC Says Use Your Own Method for Now." SEC Accounting Report 4 (October 1978): 5-6.

"SEC Says Use Your Own Method--For Now." SEC Accounting Report 4 (October 1978): 5-6.

Symonds, Edward. "That Dry Hole Business." Accountant (Eng.) 179 (November 2, 1978): 597-598.

"FASB Seeks to Shelve Rule for Standardizing Oil, Gas Accounting," Wall Street Journal, 14 November 1978, 12.

Alonzo, Martin V. "Oil and Gas Accounting: The Search for Relevance." Management Accounting 60 (November 1978): 4.

Beresford, Dennis R., and Neary, Robert D. "SEC Proposes Alternative Oil and Gas Accounting Rules." Financial Executive 46 (November 1978): 8, 10.

Bisgay, Louis. "SEC Rejects Statement No. 19." Management Accounting 60 (November 1978): 6, 60.

Randall, Robert F. "FTC Urges SEC Not to Adopt Successful Efforts Standard." Management Accounting 60 (November 1978): 5.

Randall, Robert F. "It's Rocky Going for FASB on Statements No. 8 and 19." Management Accounting 60 (November 1978): 5.

"SEC Proposes Alternative Oil and Gas Accounting Rules." Financial Executive 46 (November 1978): 8.

Norby, William C. "The SEC Decision on Oil and Gas Accounting," Financial Analysts Journal 34 (November/December 1978): 13-16, 77.

Seidler, Lee J. "Oil Companies May Have Unwanted Lifo Profits Because of Shortages." Accounting Issues (December 15, 1978): 12-13.

Davies, G. "SEC Give FASB a Slap in the Face." Accountancy (Eng.) 89 (December 1978): 13.

Seidler, Lee J. "AICPA Pronouncement Will Limit Off-Balance Sheet Financing of Airlines, Steels, Chemicals, Extractive Industries, Whiskey Companies, and Public Utilities." Accounting Issues (January 18, 1979): 18-20.

"Texaco Inc. Said It Is Revising Its Accounting Policies to Conform to Requirements of Financial Accounting Standards Board Statements No. 13 and No. 19." Wall Street Journal, 22 January 1979, 14.

"Accounting and Reporting Requirements for Oil and Gas Producers." CPA Journal 49 (January 1979): 70.

Schechtman, Daniel. "Recent SEC Developments." CPA Journal 49 (January 1979): 66-70.

"SEC Dropped Its Proposal That Would Require Oil and Gas Companies to Indicate What Their Earnings Would Be Under the Two Accounting Methods the SEC Has Approved for Such Companies on an Interim Basis." Wall Street Journal, 22 (February 1979), 7.

"Oil and Gas Rules Made to Conform More to FASB and DOE Provisions." Journal of Accountancy 147 (February 1979): 14.

"FASB Moves to Ease U.S. Oil Firms' Accounting." Oil & Gas Journal 77 (March 19, 1979): 53.

Beresford, Dennis R., and Neary, Robert D. "Oil and Gas Gets More SEC Attention." Financial Executive 47 (March 1979): 13.

Kolins, Wayne. "SEC Overrules FASB on Oil and Gas Rules." Practical Accountant 12 (March 1979); 56.

"SEC Issues Full Cost Accounting Rules and Disclosure Requirements for Oil and Gas Industry." SEC Accounting Report 5 (March 1979): 4-5.

"Go Well, Go Shell." Accountant (Eng.) 180 (April 26, 1979): 529

"Oil and Gas Producers Can Choose Between Two Accounting Methods, According to SEC Releases." Journal of Accountancy 147 (April 1979): 20, 22, 27.

"Guide Through the Maze." Accountant (Eng.) 180 (May 17, 1979): 638-639.

Beresford, Dennis R., and Neary, Robert D. "News From FASB and SEC on Oil and Gas Accounting." Financial Executive 47 (May 1979): 8.

Financial Accounting Standards Board. "Suspension of Certain Accounting Requirements for Oil and Gas Producing Companies." Statement of Financial Accounting Standards No. 25. Journal of Accountancy 147 (May 1979): 110-114.

Norby, William C. "Oil and Gas Accounting and Disclosure." Financial Analysts Journal 35 (May/June 1979): 19-20.

Symonds, Edward. "Standard Setters American Style." Accountant (Eng.) 180 (June 28, 1979): 827-828.

"FASB-Sponsored Report Evaluates Impact of Accounting Rule on Oil and Gas Companies." Journal of Accountancy 147 (June 1979): 24.

Williams, Barbara J. "Recent Developments." CPA Journal 49 (June 1979): 72-75.

"SEC Issues Full Cost Accounting Rules and Disclosure Requirements for the Oil and Gas Industry." Securities Regulation Law Journal 7 (Summer 1979): 184-186.

Beresford, Dennis R., and Neary, Robert D. "SEC Postpones an Oil and Gas Deadline." Financial Executive 47 (August 1979): 8.

"Eroding Financial Reporting Credibility." CPA Journal 50 (January 1980): 5-6.

Schechtman, Daniel. "SEC Amends and Explains Oil and Gas Accounting Rules and Delays Audit Requirement." CPA Journal 50 (March 1980): 81.

Bisgay, Louis. "MAP Comments on Oil-Gas Issue." Management Accounting 61 (June 1980): 8.

"FASB Proposal Specifies How Oil and Gas Producers Should Treat a Depletion Deduction." Journal of Accountancy 150 (September 1980): 7-8.

Williams, Barbara J. "SEC Commentary." CPA Journal 50 (September 1980): 86-89.

"FASB Proposed an Interpretation of an Accounting Rule for Exploratory Oil and Gas Wells That Turn Out to be Dry." Wall Street Journal, November 28, 1980, 16.

Bisgay, Louis. "FASB on Oil-Gas Accounting." Management Accounting 62 (November 1980): 8, 12.

Financial Accounting Standards Board. "Applying FASB Statement No. 34 to Oil and Gas Producing Operations Accounted for by the Full Cost Method." FASB Interpretation No. 33. Journal of Accountancy 150 (November 1980): 136-137.

"FASB Proposal Provides Guidance for Dry Wells." Journal of Accountancy 151 (January 1981): 7-8.

Beresford, Dennis R., and Neary, Robert D. "FASB Would Clarify Dry Hole Accounting." Financial Executive 49 (February 1981): 11.

Bisky, Thomas. "Special Report--Williams and Kirk Address AICPA's National SEC Conference." Journal of Accountancy 151 (March 1981): 7.

"SEC Commentary: Summary of 1980 Developments." CPA Journal 51 (April 1981): 68-69.

"Obscene Profits?" Wall Street Journal, 19 May 1981, 34.

Symonds, Edward. "Accounting for My Oil." Petroleum Economist (Eng.) 48 (July 1981): 289.

Meyer, Philip E. "Accounting Theory and Practice: Three Standard Setting Non-Events." Massachusetts CPA Review 55 (July/August 1981): 23-24.

Bisgay, Louis. "FASB on Oil and Gas." Management Accounting 63 (August 1981): 8.

Beresford, Dennis R., and Neary, Robert D. "Two Releases Affect Oil and Gas Producers." Financial Executive 49 (December 1981): 7-8.

Financial Accounting Standards Board. "Accounting for Exploratory Wells in Progress at the End of a Period." FASB Interpretation No. 36. Journal of Accountancy 152 (December 1981): 131-132.

Walker, Dorothy E. "ASR 261, Accounting Changes by Oil and Gas Producers Rescinded." CPA Journal 52 (February 1982): 77.

"The SEC, Investigating the Adequacy of International Metals & Petroleum Corp.'s Financial Reports, Suspended Trading in the Company's Stock for 10 days." Wall Street Journal, 30 April 1982, 7.

"RAN Energy Inc., Formerly Randado Oil & Minerals Inc. Has Further Delayed Filing Its 10K Annual Report with the SEC." Wall Street Journal, 12 May 1982, 22.

"Current Developments in Extractive Industries Accounting." Journal of Extractive Industries Accounting 1 (Spring 1982): 11-19.

Zell, Gary A. "The Relationship Between the SEC and the FASB." Ohio CPA Journal 41 (Spring 1982): 81-83.

Bisgay, Louis. "FASB Proposes Three Standards." Management Accounting 64 (July 1982): 65.

"Current Developments in Extractive Industries Accounting." Journal of Extractive Industries Accounting 1 (Summer 1982): 7-16.

Bisgay, Louis. "MAP on 'Oil and Gas' and 'R & D.'" Management Accounting 64 (Ocotber 1982): 42, 87.

"Committee Commentary on Current Developments: Proposed FAS: Oil and Gas Producing Activities Disclosures." CPA Journal 52 (November 1982): 74-79.

Symonds, Edward. "Oil and Gas Accounting--What Has the FASB Achieved?" Accountancy (Eng.) 93 (November 1982): 53-55.

Beresford, Dennis R., and Neary, Robert D. "SEC Staff Views on Various Oil and Gas Matters." Financial Executive 50 (December 1982): 12.

Bisgay, Louis. "SEC Staff on Oil, Gas, and Assets." Management Accounting 64 (December 1982): 45, 49.

"Current Developments." Journal of Extractive Industries Accounting 1 (Fall/Winter 1982): 5-6.

Securities and Exchange Commission. "Staff Accounting Bulletin 47." Journal of Extractive Industries Accounting 1 (Fall/Winter 1982): 133-157.

"The Accounting Forum." Journal of Extractive Industries Accounting 1 (Fall/Winter 1982): 7-9.

DISCLOSURES AND RESERVE RECOGNITION ACCOUNTING

"C & L Suggests New Disclosure Method for Petroleum Companies." Journal of Accountancy 143 (June 1977): 12.

"New Accounting Rule Proposed by SEC Would for First Time Require Oil and Gas Producers to Put a Dollar Value on Their Reserves." Wall Street Journal, 31 October 1977, 13.

Beresford, Dennis R., and Neary, Robert D. "SEC Wants Oil and Gas Reserves to Be Valued; Proposes to Endorse Successful Efforts Accounting." Financial Executive 45 (December 1977): 12, 14.

Bisgay, Louis. "SEC Asks for Values of Reserves." Management Accounting 59 (January 1978): 10, 60.

"SEC Decided That, Under Its Replacement Cost Accounting Procedures, Companies With Oil and Gas Reserves Needn't Estimate the Cost of Replacing Them." Wall Street Journal, 22 March 1978, 14.

Randall, Robert F. "New Replacement Cost Rule Proposed for Oil, Gas Producers." Management Accounting 59 (May 1978): 9.

Norby, William C. "Disclosure of Future Net Revenue from Oil and Gas Reserves." Financial Analysts Journal 34 (May/June 1978): 20.

"SEC Ordered the Development of a New Accounting Method for Oil and Gas Companies Based on the Value of Their Reserves." Wall Street Journal, 30 August 1978, 8.

"SEC Approves New RRA Method for Oil and Gas Accounting." Journal of Accountancy 146 (October 1978): 7-8.

"Data on Proven Reserves of Oil-Gas Firms Required in SEC's New Accounting Rules." Wall Street Journal, 21 December 1978, 10.

Weinbach, Arthur F. "Who Establishes Accounting Principles: SEC Selects New Oil and Gas Accounting Method." Connecticut CPA 42 (December 1978): 19.

Norby, William C. "Reserve Recognition Accounting for Oil and Gas Reserves." Financial Analysts Journal (January/February 1979): 10-11.

"Reserve Recognition Accounting for Oil and Gas Reserves." Financial Analysts Journal 35 (January/February 1979): 10-11.

"A Better Way to Account for Inflation?" Industry Week 200 (February 5, 1979): 98-99, 101.

"AICPA Exposes Audit Guide on Oil and Gas Reserve Data." Journal of Accountancy 147 (May 1979): 22-24.

Norby, William C. "Oil and Gas Accounting and Disclosure." Financial Analysts Journal 35 (May/June 1979): 19-20.

"SEC Decided to Give Oil and Gas Companies More Time to Come Up With Audited Estimates of Their In-The-Ground Reserves." Wall Street Journal, 14 June 1979, 16.

"Resisting the SEC Oil-Reserve Plan." Business Week (June 25, 1979): 102-103.

"Oil and Gas Industry Task Group of FASB Issues Preliminary Report on Changing Prices Disclosure." Journal of Accountancy 147 (June 1979): 24, 28.

Anreder, Steven S. "Rah-Rah: Cheers and Jeers; the SEC Has Thrust Changes on Petroleum Accounting." Barron's (August 13, 1979): 11, 18, 20-21.

Seidler, Lee J. "SEC Decides on Oil and Gas Accounting Full Cost to Continue, Current Value in the Future." Accounting Issues (August 29, 1979): 7-9.

Kolbenschlag, Michael. "Rah-Rah Gets the Bronx Cheer." Forbes 124 (September 17, 1979): 188-192.

"SEC Took Another Step Toward Its Goal of Changing the Way Oil and Gas Producers Report Their Earnings." Wall Street Journal, 20 September 1979, 4.

"Underground Profits? Oil and Gas Industry Chapes at SEC Plan to Base Accounting on Proven Reserves." Wall Street Journal, 26 September 1979, 48.

"SEC Pushes Again for Reserve Accounting." Business Week (October 1, 1979): 54, 56.

"The SEC Pushes Again for Reserve Accounting." Business Week (October 1, 1979): 54, 56.

"Price Waterhouse & Co. Attacked the SEC's Handling of Accounting for Oil-Gas Inflation." Wall Street Journal, 16 October 1979, 16.

"SEC Issues Final Rules on Oil and Gas Reporting Under RRA." Journal of Accountancy 148 (November 1979): 34.

Symonds, Edward. "How Much for Those Hidden Reserves?" Accountant (Eng.) 181 (November 1979): 624-626.

Beresford, Dennis R., and Neary, Robert D. "Another Step Toward SEC Reserve Recognition Accounting." Financial Executive 47 (December 1979): 10-11.

"SEC Orders Supplemental Oil and Gas RRA." CA Magazine (Canada) 112 (December 1979): 8.

"Analysts Say Disclosures of Reserves Will Be Useful in Their Analysis." Journal of Accountancy 149 (February 1980): 28.

"SEC Postponed a Rule That Requires Auditing of Oil and Gas Reserve Information That Companies Must File With Their Regular Financial Statements." Wall Street Journal, 17 April 1980, 4.

"New Inflation-Accounting Rules Affecting the Forest Products, Mining, Real Estate and Oil and Gas Industries Were Proposed by the Financial Accounting Standards Board, A Rule-Making Group for Accountants." Wall Street Journal, 23 April 1980, 12.

Seidler, Lee J. "Oil and Gas Accounting: A New Method to Replace Full Cost and Successful Efforts." Accounting Issues (June 16, 1980): 15.

"Inflation Accounting Guidelines for Specialized Industries." SEC Accounting Report 6 (June 1980): 6-7.

"SEC Postpones Oil and Gas Audit Rule; to Propose CPA Exemption from Liability for Soft Data." Journal of Accountancy 149 (June 1980): 8, 10.

Beresford, Dennis R., and Neary, Robert D. "SEC Postpones Audit Requirement for Oil and Gas Reserves." Financial Executive 48 (July 1980): 12.

"FASB Requires Oil and Gas, Mining Industries to Disclose Current Cost Data on Assets." Journal of Accountancy 150 (December 1980): 8, 10.

"SAS Guides Auditor's Queries About Supplementary Data on Oil and Gas Reserves." Journal of Accountancy 150 (December 1980): 7-8.

"Puzzling Petroleum: Many Investors in Oil and Gas Stocks Are Baffled by Data That Are Too Abundant and Technical." Wall Street Journal, 21 January 1981, 50.

"AICPA Standard for Supplementary Oil and Gas Information." CA Magazine (Canada) 114 (January 1981): 14.

American Institute of Certified Public Accountants. Auditing Standards Board. "Supplementary Oil and Gas Reserve Information." Statement on Auditing Standards No. 33. Journal of Accountancy 151 (January 1981): 109-110.

"FASB Calls for Further Disclosures on Assets in Resource Industries." CA Magazine (Canada) 114 (January 1981): 16.

Financial Accounting Standards Board. "Financial Reporting and Changing Prices: Specialized Assets--Mining and Oil and Gas." Statement of Financial Accounting Standards No. 39. Journal of Accountancy 151 (January 1981): 101-107.

"SEC Abandoned Its Attempt to Impose a Controversial New Accounting Method on Oil and Gas Producers and Returned Jurisdiction on the Same Issue to the Financial Accounting Standards Board." Wall Street Journal, 25 February 1981, 39.

Beresford, Dennis R., and Neary, Robert D. "FASB Addresses Inflation Disclosures for Sepcialized Industries." Financial Executive 49 (February 1981): 10.

"Financial Accounting Standards Board Will Begin Developing a Simplified System of Financial Disclosure for Oil and Gas Producers." Wall Street Journal, 5 March 1981, 52.

Baker, Thomas. "Retreat on RRA." Forbes 127 (April 27, 1981): 110.

"SEC Won't Impose RRA Method on Oil and Gas Producers; Will Support FASB Disclosure Project." Journal of Accountancy 151 (April 1981): 7-8.

Diverse Group of FASB Advisers Readies Issues Document on Oil and Gas Disclosures." Journal of Accountancy, 151 (June 1981): 7-8.

Beresford, Dennis R., and Neary, Robert D. "Oil and Gas Disclosures Revisited." Financial Executive 49 (July 1981): 10-11.

"Board Invites Comment on Oil-and-Gas Disclosures, Sets Hearing for August 18." Journal of Accountancy 152 (July 1981): 8, 10.

"FASB to Develop Oil and Gas Disclosure Standards." SEC Accounting Report 7 (July 1981): 6-7.

"Oil and Gas Disclosures Revisited." Financial Executive 49 (July 1981): 10.

"FASB Homes in on Disclosure Requirements." Oil & Gas Journal 79 (October 21, 1981): 54-55.

"SEC-Required Disclosure of Oil Reserve Values Backed at FASB Hearing." Journal of Accountancy 152 (October 1981): 14, 18.

Bisgay, Louis. "Other FASB Actions." Management Accounting 63 (January 1982): 8.

Bisgay, Louis. Other SEC News." Management Accounting 63 (January 1982): 8, 13.

Norby, William C. "Disclosures About Oil and Gas Producing Activities." Financial Analysts Journal 38 (May/June 1982): 22-23, 80.

Beresford, Dennis R., and Neary, Robert D. "FASB Proposes New Oil and Gas Disclosures." Financial Executive 50 (June 1982): 8, 10.

"Comprehensive Disclosures for Oil and Gas Producers Are Proposed by FASB." SEC Accounting Report 8 (June 1982): 6-7.

"FASB Proposal on Oil and Gas Disclosures." CA Magazine (Canada) 115 (June 1982): 16.

"Reduced Disclosure for Small Oil and Gas Producers." CPA Journal 52 (June 1982): 7. (Excerpt from Seidman & Seidman. Clients Accounting Letter, December 1981.)

Beresford, Dennis R., and Neary, Robert D. "SEC Supports FASB Oil and Gas Disclosures." Financial Executive 50 (September 1982): 18.

DRILLING FUNDS, PROGRAMS, JOINT VENTURES, AND PARTNERSHIPS

"SEC, Yielding to Oil Industry Requests, Eased Financial-Disclosure Rules for Oil Production and Other Partnerships Selling Shares to the Public." Wall Street Journal, 22 June 1982, 12.

ECONOMICS, FINANCE, AND EVALUATION

"Undermining Profits: Oil Companies' Big Investment in Mining Sours as Prices Fall, Operating Costs Rise". Wall Street Journal, 4 January 1978, 42.

"The Oil Industry's Immense Exploration and Development Costs Have Been a Source of Concern to Some Wall Street Analysts, but Look at 14 Producers' Financial Data Indicates Worries May Be Overblown." Wall Street Journal, 8 January 1979, 29.

Seidler, Lee J. "AICPA Pronouncement Will Limit Off-Balance Sheet Financing of Airlines, Steels, Chemicals, Extractive Industries, Whiskey Companies, and Public Utilities." Accounting Issues (January 18, 1979): 18-20.

"Largest Oil Concerns' Capital Spending Outstripped Earnings 53% in First Half." Wall Street Journal, 28 August 1979, 9.

"Puzzling Petroleum: Many Investors in Oil and Gas Stocks Are Baffled by Data That Are Too Abundant and Technical." Wall Street Journal, 21 January 1981, 50.

"Obscene Profits?" Wall Street Journal, 19 May 1981, 34.

"Oil Companies, Focus of Takeovers, Also Face Critical Choices in Financing Capital Outlays." Wall Street Journal, 22 July 1981, 49.

"Royalty Trusts Seen as Possible Defense for Oil Firms Worried About Takeovers." Wall Street Journal, 24 November 1981, 4.

"Breakup Values of Some Energy Firms Are Cut by John Herold, Inc., Depressing Shares of Small Oil Companies." Wall Street Journal, 9 April 1982, 35.

"The SEC, Investigating the Adequacy of International Metals & Petroleum Corp.'s Financial Reports, Suspended Trading in the Company's Stock for 10 days." Wall Street Journal, 30 April 1982, 7.

"SEC Charged Brady Energy Corp., Former Chairman Ralph Brady and Others with Securities Fraud in the Company's Sale of Preferred Stock in 1980." Wall Street Journal, 13 July 1982, 16.

"SEC Sued to Stop the Sale of Securities by Telex Gas & Oil Co., Charging That Its Executives Violated Securities Laws." Wall Street Journal 25 August 1982, 4.

Stock, F. "Energy Financing; Cautious Way Forward for Banks." Petroleum Economist (Eng.) 49 (November 1982): 445-446.

INTERNATIONAL

Clutterbuck, D. "Pertamina Puts Oil Project First." International Management 33 (October 1978): 34-38.

Accounting Standards Committee. "Accounting for Petroleum Revenue Tax." Proposed Statement of Standard Accounting Practice, ED28. Management Accounting (Eng.) 59 (May 1981): 68-70.

"CICA Releases Proposed Oil and Gas Guideline." CA Magazine (Canada) 114 (June 1981): 16.

"Accounting Guidelines on PIP and PGRT Released." CA Magazine (Canada) 115 (March 1982): 9.

ELECTRONIC DATA PROCESSING

"Combining COM with an Online System to Speed Access." Information & Records Management 15 (July 1981): 32-34.

"News Minis Update Chevron's Old CPUs." Computerworld 15 (August 3, 1981): 51.

AUDITING

"AICPA Exposes Audit Guide on Oil and Gas Reserve Data." Journal of Accountancy 147 (May 1979): 22-24.

"SEC Decided to Give Oil and Gas Companies More Time to Come Up With Audited Estimates of Their In-The-Ground Reserves." Wall Street Journal, 14 June 1979, 16.

Schechtman, Daniel. "SEC Amends and Explains Oil and Gas Accounting Rules and Delays Audit Requirement." CPA Journal 50 (March 1980): 81.

"SEC Postponed a Rule That Requires Auditing of Oil and Gas Reserve Information That Companies Must File With Their Regular Financial Statements." Wall Street Journal, 17 April 1980, 4.

"SEC Postpones Oil and Gas Audit Rule; to Propose CPA Exemption from Liability for Soft Data." Journal of Accountancy 149 (June 1980): 8, 10.

Beresford, Dennis R., and Neary, Robert D. "SEC Postpones Audit Requirement for Oil and Gas Reserves." Financial Executive 48 (July 1980): 12.

Bisgay, Louis. "SEC Postpones Oil and Gas Audit Rules." Management Accounting 62 (July 1980): 10, 59.

"SAS Guides Auditor's Queries About Supplementary Data on Oil and Gas Reserves." Journal of Accountancy 150 (December 1980): 7-8.

American Institute of Certified Public Accountants. Auditing Standards Board. "Supplementary Oil and Gas Reserve Information." Statement on Auditing Standards No. 33. Journal of Accountancy 151 (January 1981): 109-110.

"GAO Hits Auditing of Tertiary Incentive Program." Oil & Gas Journal 79 (November 1981): 423.

GOVERNMENT REGULATION AND PRICE CONTROL

Seidler, Lee. J. "Department of Energy Hearings on Oil and Gas Accounting." Accounting Issues (March 14, 1977): 10-13.

"DOE Holds Hearings on Oil and Gas Accounting." Journal of Accountancy 145 (April 1978): 10.

"The Energy and Justice Departments Have Drafted a Broad Lawsuit Accusing Most of the Largest U.S. Oil Companies of Overcharging Their Customers at Least $1 Billion for Certain Petroleum Products." Wall Street Journal, 5 January 1979, 5.

"Officials of Nine of the Nation's Largest Oil Companies Reacted Bitterly to the Government's Massive Lawsuit Alleging at Least $1 Billion in Overcharges for Natural Gas Liquids. They Denied the Allegations and Accused Energy Dept. of Retroactive Reinterpretations of Complex Petroleum Pricing Regulations That Stemmed from the 1973-74 Arab Oil Embargo." Wall Street Journal, 8 January 1979, 8.

"The Energy Dept. Is Hiring an Independent Auditor to Check the Reliability of Petroleum Inventory Data Reported by the Oil Companies to the Government." Wall Street Journal, 11 June 1979, 2.

Winkler, Connie. "U.S. Losing Out on Gas Royalties." Computerworld 13 (July 2, 1979): 1, 6.

"Cost, Price Data of Oil Companies Sought Quickly; Wage Panel to Determine if 20 Firms Are Violating Anti-Inflation Standards." Wall Street Journal, 29 October 1979, 5.

"Data of Five Big Oil Concerns Is Held Reliable by Audit, Which Faults Questions of U.S." Wall Street Journal, 29 October 1979, 5.

"Department of Energy MAPS Industry Path in Wake of Decontrol." Oil & Gas Journal 79 (February 23, 1981): 54.

"GAO Hits Auditing of Tertiary Incentive Program." Oil & Gas Journal 79 (November 1981): 423.

"SEC Charged Brady Energy Corp., Former Chairman Ralph Brady and Others with Securities Fraud in the Company's Sale of Preferred Stock in 1980." Wall Street Journal, 13 July 1982, 16.

"SEC Sued to Stop the Sale of Securities by Telex Gas & Oil Co., Charging That Its Executives Violated Securities Laws." Wall Street Journal 25 August 1982, 4.

PIPELINING AND TRANSMISSION

"Apportionment of Costs." Public Utilities Fortnightly 100 (November 10, 1977): 47-49.

"Cost Apportionment is Key in Gas Rate Cases." Public Utilities Fortnightly 99 (February 17, 1977): 48-49.

"Cost Classification and Allocation Methods in Gas Rates." Public Utilities Fortnightly 99 (June 23, 1977): 52_53.

Federal Energy Regulatory Commission Proposed Requiring Producers of Alaskan Natural Gas to Pay for Processing It, Rather Than a Pipeline Company That Would Pass the Cost to Buyers." Wall Street Journal, 5 February 1979, 4.

"Rate-Making Treatment Set for Gas Exploration Costs." Public Utilities Fortnightly 108 (July 16, 1981): 61.

INTEREST CAPITALIZATION

"Financial Accounting Standards Board Appointed 16-Member Advisory Panel to Consider How Companies Should Account for Interest Costs." Wall Street Journal, 1 October 1975, 3.

"FASB Issued Memorandum on How Interest Costs Should be Treated." Wall Street Journal, 22, December 1977, 4.

"FASB Narrowly Clears Rule on How Companies Handle Interest Costs." Wall Street Journal, 23, October 1979, 21.

Beresford, Dennis R., and Neary, Robert D. "FASB Explains Interest Capitalization for Oil and Gas Producers." Financial Executive 48 (October 1980): 10.

DEBT EXTINGUISHMENT

"Debt-Reduction Method Draws Concern of SEC; Staff Expected to Seek Halt to 'Defeasance,' Used by Exxon Corp and Kellog Co." Wall Street Journal, 13 August 1982, 5.

"The SEC Bans the Use of 'Defeasance' Technique, a Controversial Accounting Method Used to Spruce Up the Financial Statements of Publicly Held Companies, Until FASB Decision." Wall Street Journal, 20 August 1982, 7.

Beresford, Dennis R., and Neary, Robert D. "'Defeasance' Practices Halted Pending FASB Study." Financial Executive 50 (October 1982): 10.

Beresford, Dennis R., and Neary, Robert D. "Debt Extinguishment Problems Get FASB's Attention." Financial Executive 50 (December 1982): 12.

OIL AND GAS
BOOKS
1973-1982

Allain, Louis J. Capital Investment Models of the Oil & Gas Industry: A Systems Approach. Energy in the American Economy Series. Stuart Bruchey, ed. Salem, N.H.: Ayer Co., 1979.

Barrett, M. Edgar, and Cormack, Mary Pat. Management Strategy in the Oil and Gas Industries: Cases and Readings. Houston: Gulf Publishing Company, 1982.

Bedingfield, James. Accounting & Federal Regulation. Reston, Va.: Reston Publishing Co., 1982.

Brock, Horace R.; Klingstedt, John P.; and Jones, Donald M. Accounting for Oil & Gas Producing Companies--Part 1: Exploration, Acquisition, Development and Production. Denton, Tex.: Professional Development Institute, 1981.

Brock, Horace R.; Klingstedt, John P.; and Jones, Donald M. Accounting for Oil & Gas Producing Companies--Part 2: Amortization, Conveyances, Full Costing and Disclosures. Denton, Tex.: Professional Development Institute, 1982.

Council on Economic Priorities, and Boothe, Norris. Cleaning Up: The Cost of Refinery Pollution Control. New York: Council on Economic Priorities, 1975.

Crumbley, Larry D., and Grossman, Steven D. Readings in Oil Industry Accounting. Tulsa: Penn Well Publishing Company, 1980.

de Reyna, Ramon J. "Recent Significant Developments Affecting Accounting and Reporting Practices of Oil and Gas Exploration Companies," in Annual Accounting Review, New York; 1980, vol. 2, 127-141.

Hardy, George, and Arnold, Carol. Royalties & Division Orders. Oklahoma City: Institute for Energy Development, 1981.

Hardy, George. Regulation of Oil & Gas Profits. Oklahoma City: Institute for Energy Development, 1981.

Hass, Jerome E.; Mitchell, Edward J.; and Stone, Bernell K. Financing the Energy Industry. Cambridge, Mass.: Ballinger Publishing Company, 1974.

Horngren, Charles T. "Politics and Accounting Policymaking," in Al-Hashim, Dhia D., and Robertson, James W. Contemporary Issues in Accounting. Indianapolis: Bobbs-Merrill Co., 1979, 208-221.

Koester, Robert. Advanced Oil & Gas Accounting. Oklahoma City: Institute for Energy Development, 1981.

Koester, Robert. Basic Oil & Gas Accounting. Oklahoma City: Institute for Energy Development, 1981.

Mosburg, Lewis G. Letter of Credit Oil Programs. Oklahoma City: Institute for Energy Development, 1979.

Mosburg, Lewis G., Jr. Contract Use in Oil & Gas Operations. Oklahoma City: Institute for Energy Development, 1981.

Mosburg, Lewis G., Jr. Profiting from the Energy Crisis. Oklahoma City: Institute for Energy Development, 1982.

Myers, John H. Full Cost vs Successful Efforts in Petroleum Accounting: An Empirical Approach. Ad Hoc Committee (Petroleum Companies) on Full Cost Accounting. Indiana University Press, 1974.

Nelson, W. L. Guide to Refinery Operating Costs (Process Costimating). 3rd ed. Tulsa: Petroleum Publishing Company, 1976.

Norton, John C., and Rowe, Donald A. Accounting and Auditing Guide for United Kingdom Oil and Gas Exploration and Production. Cambridge: Heffers Printers.

Owen, Thomas L. "Oil and Gas Industries," in Levine, Sumner N. Financial Analyst's Handbook. Homewood, Ill., 1975, v. 2, 544-585.

Page, John S. Cost Estimating Manual for Pipelines and Marine Structures. Houston: Gulf Publishing Company, 1977.

Petroleum Analysis Limited. Reporting Financial and Operating Data on Oil and Gas Producing Activities of Petroleum Companies. New York.

"Petroleum Production," in Encyclopedia of Accounting Systems. Rev. ed. Englewood Cliffs, N.J., 1976, 1383-1416.

Redford, Emmette S., ed. Public Administration & Policy Formation: Studies in Oil, Gas, Banking, River Development, & Corporate Investigations. Westport, Conn.: Greenwood Press. reprint of 1956 edition.

Sheets, Robert. Market Value Royalties. Oklahoma City: Institute for Energy Development, 1981.

Tellini, William P. "Oil, Gas, and Other Natural Resources," in Seidler, Lee Jr., and Carmichael, Douglas R. Accountants' Handbook, 6th ed. New York: Ronald Press, 1981, v. 1, sect. 19.

Whitman, Robert O. "Current Problems and Practical Solutions of Accounting and Reporting for Regulated Electric and Gas Utilities," in Edwards, James Don. Modern Accountant's Handbook. Homewood, Ill., 1976, 917-946.

OIL AND GAS
BOOKLETS, REPORTS, RESEARCH STUDIES
1973-1982

Adkerson, Richard C. Reserve Recognition Accounting Revisited: A Review of 1979 Oil and Gas Disclosures by Public Companies. Chicago: Arthur Andersen & Company, 1981.

American Accounting Association. Committee on Financial Accounting Standards. Subcommittee on Accounting for the Extractive Industries. Report to the Financial Accounting Standards Board. Sarasota, Fla., 1977.

American Petroleum Institute. Report on Certain Petroleum Industry Accounting Practices. Washington, 1974.

Arthur Andersen & Company. Accounting for Income Taxes in the United States. Chicago, 1977.

Arthur Andersen & Company. Financial Accounting and Reporting by Oil and Gas Producing Companies: Response to the FASB on Its Proposed Statement. Chicago, 1977.

Arthur Young & Company. Successful Efforts Accounting: Why It Is Needed in the Extractive Industries. New York, 1977.

Arthur Young & Company. Financial Accounting and Reporting in the Extractive Industries: Comments, to the Financial Accounting Standards Board. New York, March 1977.

Arthur Andersen and Company. Accounting for Petroleum Products Futures Contracts. Chicago, 1982.

Baggett, Monte R.; Cole, Richard D,; and Short, Jack E. Anatomy of a Drilling Fund. New York: Coopers & Lybrand, 1980.

Bybee, W. Cecil. "Valuation Accounting for Oil and Gas Viewpoint," in Viewpoint. 2nd ed. Main Lafrentz & Company, 1978, 7-9.

Chase Manhattan Bank. Energy Economics Division. Financial Analysis of a Group of Petroleum Companies. New York, 1971, 1974-1977, 1979.

Collins, Daniel, and Dent, Warren. "An Empirical Assessment of the Stock Market Effects of the Proposed Elimination on Full Cost Accounting in the Extractive Petroleum Industries." University of Iowa, 1978.

Continental Oil Company. Oil, Coal, and Uranium: Do They Mix? Stamford, Conn., 1977.

Coopers & Lybrand. Planning for Energy Compliance. New York, 1980.

Deloitte Haskins & Sells. Accounting Practices 1978--Illustrative Items of Current Interest from Published Annual Reports. New York, 1979.

Dunn, James F. "Tribulations of a Proposed APB Opinion on the Extractive Industries," in Haskins & Sells Selected Papers, 1972. New York, 1973, 98-104.

Grier, J. D., et al.. Financial Accounting for Oil and Gas Reserves. New York: Peat, Marwick, Mitchell & Company, 1980.

Haskins & Sells. Financial Accounting and Reporting by Oil and Gas Producing Companies: Guidelines for Application of FASB Statement No. 19. New York, 1978.

Hunt, John W. "Accounting for Oil and Gas," in Viewpoint. 2nd ed. Main Lafrentz & Company, 1978, 2-6.

Lev, Baruch. Impact of Accounting Regulation on the Stock Market: The Case of Oil and Gas Companies. Chicago: University of Chicago, Center for Mathematical Studies in Business and Economics, 1978.

Los Alamos Scientific Laboratory. Accounting for New Profit Share Leases. May 1981.

May, Alan, Jr. The Revised Petroleum Accounting Rules. New York: Coopers & Lybrand, 1980.

Most, Kenneth S. Financial Reporting by the Oil and Gas Industry in Europe: A Survey Based on 1974 Annual Reports. Miami: Florida International University, 1976.

Most, Kenneth S. Financial Reporting by the Oil and Gas Industry in Europe: A Survey Based on 1975 Annual Reports. Miami: Florida International University, 1977.

Most, Kenneth S., and Mulford, Charles W. Financial Reporting by the Oil and Gas Industry in Europe: A Survey Based on 1977 Annual Reports. Miami: Florida International University, 1979.

North American Securities Administrators Association. Oil and Gas Interests Committee. Draft of Revisions of Guidelines for the Registration of Oil and Gas Programs. Chicago: Commerce Clearing House, 1976.

O'Connor, Melvin, and Collins, Daniel. "Full Cost vs Successful Efforts Accounting in the Oil and Gas Industries: A Closer Look at the Potential Market Consequences." Michigan State University, 1977.

Peat, Marwick, Mitchell & Company. Financial Accounting and Reporting by Oil and Gas Producing Companies. New York, September 22, 1977.

Peat, Marwick, Mitchell & Company. Financial Accounting and Reporting in the Extractive Industries. New York, March 7, 1977.

Peat, Marwick, Mitchell & Company. Financial Accounting for Oil and Gas Reserves. Houston, 1979.

Peat, Marwick, Mitchell & Company. Financial Accounting for Oil and Gas Reserves. New York, 1979.

Peat, Marwick, Mitchell & Company. Principles and Presentation: Oil and Gas. New York, 1976.

Peat, Marwick, Mitchell & Company. Principles and Presentation: Oil and Gas. New York, 1977.

Peat, Marwick, Mitchell & Company. Principles and Presentation: Oil and Gas. New York, 1978.

Peat, Marwick, Mitchell & Company. Principles and Presentation: Oil and Gas. New York, 1981.

Peat, Marwick, Mitchell & Company. Principles and Presentation: Oil and Gas. New York, 1982.

Peat, Marwick, Mitchell & Company. Principles and Presentation: Refining. New York, 1981.

Peat, Marwick, Mitchell & Company. Summary Results of 1979 Survey of Members of National Association of Petroleum Investment Analysts Regarding Disclosure of Oil and Gas Quantities and Values, Supplemental Earnings, Summary, and Reserve Recognition Accounting. New York, 1979.

Peat, Marwick, Mitchell & Company. 1979 Survey by Peat Marwick Mitchell of Petroleum Investment Analysts. New York, 1979.

Porter, Stanley P. Highlights of a Study of the Subjectivity of Reserve Estimates and its Relation to Financial Reporting. New York: Arthur Young & Company, 1980.

Price Waterhouse & Company. Accounting Controls in the Petroleum Industry: Summary Guide for Management and Directors. New York, 1979.

Price Waterhouse & Company. Oil and Gas Industry in Canada: 1977 Survey of Financial Reporting and Accounting Developments. New York, 1977.

Price Waterhouse & Company. Oil and Gas Industry in Canada: 1978 Survey of Financial Reporting and Accounting Developments. New York, 1978.

Price Waterhouse & Company. Reserve Recognition Accounting: An Evaluation of its Viability and Application. New York, 1979.

Price Waterhouse & Company. Survey of Financial Reporting and Accounting Developments in the Petroleum Industry, Based on 1972 Reporting. New York, 1973.

Price Waterhouse & Company. Survey of Financial Reporting and Accounting Developments in the Petroleum Industry Based on 1973 Reporting. New York, 1974.

Price Waterhouse & Company. Survey of Financial Reporting and Accounting Developments in the Petroleum Industry, Based on 1975 Reporting. New York, 1976.

Price Waterhouse & Company. Survey of Financial Reporting and Accounting Developments in the Petroleum Industry Based on 1976 Reporting. New York, 1977.

Price Waterhouse & Company. Survey of Financial Reporting and Accounting Developments in the Petroleum Industry Based on 1977 Reporting. New York, 1978.

Price Waterhouse & Company. Survey of Financial Reporting and Accounting Developments--The Oil and Gas Industry in Canada, 1977. New York, 1977.

Society of Petroleum Engineers. Standards Pertaining to the Estimating and Auditing of Oil and Gas Reserve Information. Dallas, 1980. Reprint from Journal of Petroleum Technology (December 1979): 1559-1565.

Stone, Richard H. "Oil and Gas Investment: Tax and Economic Considerations," in Viewpoint. 2nd ed. Main Lafrentz & Company, 1973, 58-63.

Touche Ross & Company. Oil and Gas Accounting. New York, 1979.

Touche Ross & Company. Oil and Gas Accounting. New York, 1980.

Welsch, Glenn A., and Deakin, Edward B. Measuring and Reporting the "Replacement" Cost of Oil and Gas Reserves: A Research Study. Washington, D.C.: API, 1977.

Zecher, J. Richard. Full Cost vs. Successful Efforts: A Study of a Proposed Accounting Change's Competitive Impact. SEC Directorate of Economic and Policy Research, February, 1978.

OIL AND GAS
ARTICLES
1973-1982

GENERAL

"Accounting for Oil." Accountant (Eng.) 168 (March 1973): 279.

Adkerson, Richard C. "Accounting for Oil and Gas Exchange Offers." Journal of Extractive Industries Accounting 1 (Spring 1982): 21-40.

Alkire, Marion H.; Mersenheimer, F. E.; and Carroll, B. Q. "Accounting for Net Profits Interests in Oil and Gas Producing Operations." Journal of Extractive Industries Accounting 1 (Summer 1982): 67-72.

Anderson, Mark E., and Ratcliffe, Thomas A. "Accounting in Oil and Gas Industry: Examination of SFAS No. 19." Oil & Gas Tax Quarterly 27 (June 1979): 444-459.

Arnold, Jerry L., and Diamond, Michael A. "Accounting for Oil and Gas: The FASB Faces Political Realities." California CPA Quarterly 47 (June 1979): 29-32.

Baker, C. Richard. "Defects in Full-Cost Accounting in the Petroleum Industry." Abacus (Australia) 12 (December 1976): 152-158.

Baker, C. Richard. "Discussion of Drawbacks to Full-Cost in Petroleum Industry Accounting." Oil & Gas Tax Quarterly 25 (March 1977): 306-319.

Barton, M. Frank, Jr. "Allocating Costs of Joint Products in the Petroleum Industry." Oil & Gas Tax Quarterly 29 (March 1981): 578-594.

Benjamin, Wahjudi P., and McEnroe, John E. "FASB's Policy Intervention and the Behavior of Security Returns: An Empirical Investigation of the Oil and Gas Industry Employing a Combination of the ITS and the SUR Models." Journal of Business Finance & Accounting 8 (Autumn 1981): 303-327.

Bhat, S. N. "Special Problems of Oil and Gas Accounting." Management Accountant (India) 17 (May 1982): 257-259.

Bierman, Harold; Dukes, Ronald E.; and Dyckman, Thomas R. "Financial Accounting in the Petroleum Industry." Journal of Accountancy 138 (October 1974): 58-64.

Brock, Horace R.; Klingstedt, John P.; and Jones, Donald M. "General Guide to the Full-Cost Accounting Method." Journal of Extractive Industries Accounting 1 (Summer 1982): 17-35.

Brown, W. Wayne, and Grove, Hugh D. "Another Compromise Solution: A Critical Analysis of SFAS No. 19." Oil & Gas Tax Quarterly 26 (June 1978): 415-435.

Burrow, Jackie. "Treatment of Dry Holes Under the Successful-Efforts Method of Accounting--A Definitional Problem." Journal of Extractive Industries Accounting 1 (Spring 1982): 57-62.

"Can a Dry Hole Be an Asset?" Forbes 119 (May 15, 1977): 129, 131.

Clemison, John. "Oilmen." Accountant (Eng.) 184 (March 26, 1981): 331-333.

Collins, Daniel W.; Dent, Warren T.; and O'Connor, Melvin C. "Market Effects of the Elimination of Full Cost Accounting in the Oil and Gas Industry." Financial Analysts Journal 34 (November/December 1978): 48-57.

Collins, Daniel W., and Dent, Warren T. "Proposed Elimination of Full Cost Accounting in the Extractive Petroleum Industry: An Empirical Assessment of the Market Consequences." Journal of Accounting and Economics (Netherlands) 1 (March 1979): 3-44.

Collins, Daniel W., and O'Connor, Melvin C. "Examination of the Association Between Accounting and Share Price Data in the Extractive Petroleum Industry: A Comment and Extension." Accounting Review 53 (January 1978): 228-239.

Collins, Daniel W.; Rozeff, Michael S.; and Dhaliwal, Dan S. "Economic Determinants of the Market Reaction to Proposed Mandatory Accounting Changes in the Oil and Gas Industry: A Cross-Sectional Analysis." Journal of Accounting & Economics (Netherlands) 3 (March 1981): 37-71.

Collins, Daniel W.; Rozeff, Michael S.; and Salatka, William K. "SEC's Rejection of SFAS No. 19: Tests of Market Price Reversal." Accounting Review 57 (January 1982): 1-17.

Connor, Joseph E. "Putting Accounting Developments in Perspective." Financial Executive 42 (May 1974): 74-78, 80.

Cooper, Kerry; Flory, Steven M.; and Grossman, Steven D. "New Ball Game for Oil and Gas Accounting." CPA Journal 49 (January 1979): 11, 13-14, 16-17.

Cowperthwaite, G. H. "Financial Reporting: Applications and Interpretations." Accountant (Eng.) 178 (April 13, 1978): 488-490.

Deakin, Edward B. "Analysis of Differences Between Non-Major Oil Firms Using Successful Efforts and Full Cost Methods." Accounting Review 54 (October 1979): 722-734.

Deakin, Edward B. "Auditor Selection, Organization Control, Adverse Events, and the Selection of Accounting Method for Oil Exploration." Quarterly Review of Economics and Business, 20 (Autumn 1980): 77-85.

De Haven, Donald L., and De Haven, Carol S. "AICPA and FASB: Environs and Decisions." Woman CPA 40 (July 1978): 28-30.

Demers, Louis, and Wayland, Donald. "Corporate Social Responsiblity: Is No News Good News?--Part 2." CA Magazine (Canada) 115 (February 1982): 56-60.

Deview, B. J. "Accounting for the Extractive Industries." Australian Accountant 46 (October 1976): 549-550, 555.

Dopuch, Nicholas, and Sunder, Shyam. "FASB's Statements on Objectives and Elements of Financial Accounting: A Review." Accounting Review 55 (January 1980): 1-21.

Dougherty, John. "Historical Cost System Still the Best." CA Magazine (Canada) 110 (September 1977): 67-69.

DuBois, Donald A. "Full Costing in the Petroleum Industry a Test of the Efficient Market Hypothesis." Accounting Journal 1 (April 1977): 103-114.

Durand, Francis L. "Financial Reporting Effects of Changed Percentage Depletion Allowance in Oil and Gas Industry." Oil & Gas Tax Quarterly 24 (September 1975): 65-126.

Dyckman, Thomas R. "Market Effects of the Elimination of Full-Cost Accounting in the Oil and Gas Industry: Another View." Financial Analysts Journal 35 (May/June 1979): 75-80.

Dyckman, Thomas R., and Smith, Abbie J. "Financial Accounting and Reporting by Oil and Gas Producing Companies: A Study of Information Effects." Journal of Accounting and Economics (Netherlands) 1 (March 1979): 45-75.

Eskew, Robert K. "An Examination of the Association Between Accounting and Share Price Data in the Extractive Petroleum Industry." Accounting Review 50 (April 1975): 316-324.

Eskew, Robert K. "An Examination of the Association Between Accounting and Share Price Data in the Extractive Industry: A Reply." Accounting Review 53 (January 1978): 240-246.

Esposito, P., and Carter, J. "Accounting for the Oil and Gas Industry--Its Past, Present, and Future." The GAO Review 11 (Winter 1977): 57-69.

Etabari, Ahmad. "The Reaction of Oil Company Share Prices to the Outbreak of the Iranian-Iraqi War." Journal of Extractive Industries Accounting 1 (Fall/Winter 1982): 117-123.

Extractive Industries Accounting Research Institute. "Survey of Successful-Efforts Accounting Techniques." Journal of Extractive Industries Accounting 1 (Summer 1982): 91-119.

Extractive Industries Accounting Research Institute. "A Survey of Full-Cost Accounting Practices." Journal of Extractive Industries Accounting 1 (Fall/Winter 1982): 83-97.

Feller, Robert E. "Accounting for Joint Products in the Petroleum Industry." Management Accounting 59 (September 1977): 41-44, 48.

Ferguson, Robert, and Popkin, Philip. "Pulling Rabbits Out of Hats in the Oil Business--and Elsewhere." Financial Analysts Journal 38 (March/April 1982): 24-27.

Field, Robert E. "Financial Reporting in the Oil Industry." Price Waterhouse Review 19, No. 1 (1974): 4-13.

Flory, Steven M., and Grossman, Steven D. "New Oil and Gas Accounting Requirements." CPA Journal 48 (May 1978): 39-43.

Flory, Steven, M., and Grossman, Steven D. "Oil and Gas Accounting Controversy." Louisiana Certified Public Accountant 37 (Summer/Fall 1978): 27-35.

Foster, George. "Accounting Policy Decisions and Capital Market Research." Journal of Accounting and Economics (Netherlands) 2 (March 1980): 29-62.

Gaertner, James F. "FASB Statements 2 and 19: Recent SEC Action and Possible Repercussion." University of Michigan Business Review 31 (September 1979): 8-11.

Giroux, Gary A.; Kratchman, Stanley H.; and Grossman, Steven D. "Inflation Accounting for the Oil and Gas Industry." Oil & Gas Tax Quarterly 30 (December 1981): 324-340.

Grier, James D., and Moore, Cecil H. ("C. H."). "Accounting for Deferred Income Taxes of Oil and Gas Producers." Journal of Extractive Industries Accounting 1 (Spring 1982): 41-56.

Hagerman, Robert L. "Full Costing in the Petroleum Industry, A Test of the Efficient Market Hypothesis: A Critique." Accounting Journal 1 (Spring 1977): 115-116.

Heazlewood, C. T. "Accounting for the Extractive Industry." Chartered Accountant in Australia 45 (September 1974): 6-9.

Herring, Hartwell C., and Jacobs, Fred A. "Are Oil Companies the Robber Barons of the 20th Century?" Business 29 (November/December 1979): 10-15.

Hillison, William, and Dunn, Marcus. "Oil and Gas Producers Required to Use Current Value Accounting?" Oil & Gas Tax Quarterly 27 (June 1979): 460-474.

Horwitz, Bertrand, and Kolodny, Richard. "Financial Reporting Regulation and Small Research-Intensive Firms." Journal of Small Business Management 20 (January 1982): 44-49.

"How Oil Company Shareholders Took the Energy Crisis." Institutional Investor 8 (July 1974): 41.

Hunt, John W. "Accounting for Oil and Gas." Viewpoint (1978): 2-6.

Ijri, Yuji. "Oil and Gas Accounting: Turbulence in Financial Reporting." Financial Executive 47 (August 1979): 18-26.

Kelley, Robert. "Accounting for the Effects of Reserve Changes on Interim Results." Journal of Extractive Industries Accounting 1 (Fall/Winter 1982): 11-19.

Khumawala, Shaleha B.; Moylan, James E.; and Pratt, James W. "Accounting and Tax Treatment for Offshore Production Platform Removal Costs." Oil & Gas Tax Quarterly 30 (September 1981): 95-106.

King, Barry G. "Impairment Test for Unproved Properties." Journal of Extractive Industries Accounting 1 (Summer 1982): 79-89.

Koester, Robert J.; Munter, Paul; and Ratcliffe, Thomas A. "The Real Problem in Oil and Gas Accounting." Oil and Gas Tax Quarterly 30 (September 1981): 146-160.

Kolton, Paul. "The FASB in the 1980s: Standard Setting in a Changing Environment." Journal of Accountancy 153 (March 1982): 84-92.

Krasts, Aivars. "It's Time to Break with Tradition." Public Relations Journal 36 (April 1980): 54-57.

Kross, William. "Stock Returns and Oil and Gas Pronouncements: Replication and Extension." Journal of Accounting Research 20 (Autumn 1982): 459-471.

Lay, David W. "Petroleum Industry Accounting Reform on the Way." CA Magazine (Canada) 110 (December 1977): 31-34.

Lev, Baruch. "The Impact of Accounting Regulation on the Stock Market: The Case of Oil and Gas Companies." Accounting Review 54 (July 1979): 485-503.

Lilien, Steven, and Pastena, Victor. "Intramethod Comparability: The Case of the Oil and Gas Industry." Accounting Review 56 (July 1981): 690-703.

Loomis, Carol J. "How to Think About Oil-Company Profits." Fortune 89 (April 1974): 98-103, 192, 194, 196, 198.

Lourens, Roy. "Industry Attitudes to Accounting Research Study No. 11." Chartered Accountant in Australia 43 (January 1973): 20-23.

Lyon, Robert A. "U.S. Oil Accounting: One Man's View." Accountant's Magazine (Scot.) 79 (November 1975): 386-388.

Marcus, P. C. "Outline of Oil Exploration and Costing." Singapore Accountant 12 (1977): 79-82.

Matteotti, Dennis A., and Gebicke, Mark E. "Energy Policy and Conservation Act of 1975 and Its Implications for GAO." GAO Review 11 (Summer 1976): 37-42.

Meyer, Philip E. "Accounting for Oil and Gas Exploration Costs." Massachusetts CPA Review 52 (March/April 1978): 26-27.

Meyer, Philip E. "Oil and Gas Accounting--The Sequel." Massachusetts CPA Review 52 (November/December 1978): 31.

Morris, R. C. "Comparability of Oil Company Accounts: A Comment." Accounting and Business Research (Eng.) 6 (Winter 1975): 70-78.

Most, Kenneth S. "A Comparative Study of the Accounts of Seven Major Oil Companies." Accounting and Business Research (Eng.) (Autumn 1974): 242-250.

Most, Kenneth S. "Oil Company Accounting: Not So Comparable? A Reply." Accounting and Business Research (Eng.) 6 (Winter 1975): 67-69.

Nagger, Ali. "Oil and Gas Accounting: Where Wall Street Stands." Journal of Accountancy 146 (September 1978): 72-77.

Nethercott, Leslie J. "Reporting in the Petroleum Industry Under Full Cost Accounting." Accountant's Magazine (Scot.) 80 (April 1976): 131-133.

Noonan, Edward R. "1973 Accounting Principles--Evolution or Revolution." Financial Forum 11 (April 1973): 1-5. (Reprinted from The Journal of Commercial Bank Lending, January 1973).

Norr, David. "Some Elements of Oil Profitability." Financial Analysts Journal 29 (November/December 1973): 58-66.

O'Connor, Melvin C., and Collins, Daniel W. "Controversy and Research Related to Proposed Elimination of Full-Cost Accounting in Oil and Gas Industries." Oil & Gas Tax Quarterly 27 (September 1978): 54-66.

"Our Oil and Gas Practice: Dramatic Growth and an Exciting Future." People of Arthur Andersen 9 (No. 2, 1982): 20-25, 34-38.

Oxner, Thomas, H. "Development of Segmental Reporting in Oil and Gas Industry." Oil & Gas Tax Quarterly 27 (September 1978): 67-79.

Peppers, D. A., and Wellman, M. R. "How New Accounting Methods Affect Production Payments." World Oil 186 (May 1978): 99-100.

Ratcliffe, Thomas A.; Anderson, Mark E.; and Munter, Paul. "On the Development of Accounting in the Oil and Gas Industry." Oil & Gas Tax Quarterly 29 (December 1980): 349-363.

"Rate-Making Treatment Set for Gas Exploration Costs." Public Utilities Fortnightly 108 (July 16, 1981): 61.

Reed, Joel L. "Exploring for Information of Oil and Gas Companies." Financial Analysts Journal 34 (November-December 1978): 42-46.

Reilly, Ann M. "Caught Between Conflicting Regulations." Dun's Review 113 (April 1979): 52-54, 58.

Ridenour, P. Dean. "Special Problems in Computing Amortization of Capitalized Costs Under Full Costing." Journal of Extractive Industries Accounting 1 (Summer 1982): 37-45.

Schnepper, Jeff A. "Financial and Tax Accounting in the Oil Industry." National Public Accountant 21 (July 1976): 16-23.

Seidler, Lee J. "FASB Oil and Gas Draft Will Raise Future Earnings of MESA Petroleum." Accounting Issues (September 15, 1977): 4-16.

Seidler, Lee J. "Impact of New FASB Oil and Gas Accounting Rule on Five Companies." Accounting Issues (November 8, 1977): 5-11.

Seidler, Lee J. "Major Points in the FASB Oil and Gas Exposure Draft." Accounting Issues (August 15, 1977): 4-13.

Seidler, Lee J. "Some Industrial Companies Will Be Sharply Affected by the FASB's Oil and Gas Proposals." Accounting Issues (December 7, 1977): 12-27.

Shakin, Bernard. "Fight Like Wildcats Oil Independents Mount a Drive Against Accounting Change." Barron's (July 10, 1978): 11-12, 18.

Shearon, Winston T., and Kratchman, Stanley H. "Alternative Cash-Flow Analysis for Major Oil and Gas Producing Companies." Oil & Gas Tax Quarterly 28 (June 1980): 538-548.

Smartt, Lucien E. "How a Law May Work Against the Congressional Purpose." Public Utilities Fortnightly 100 (October 13, 1977): 6-8.

Smith, Abbie, and Dyckman, Thomas. "The Impact of Accounting Regulation on the Stock Market: The Case of Oil and Gas Companies: A Comment." Accounting Review 56 (October 1981): 959-966.

Smith, A. J. "SEC Reversal of FASB Statement No. 19: An Investigation of Information Effects." Journal of Accounting Research 19 (1982 Supplement): 174-226.

Smith, Alan F. "Oil Production and Accounting in the Extractive Industry." Abacus (Australia), 10 (December 1974): 100-110.

Smith, David D. "Accounting for Holes in the Ground." Accountant's Magazine (Scot.) 78 (May 1974): 171-174.

Smith, Philip S., and Herrmann, Michael C. "Ceiling Test for Full Cost Companies." Journal of Extractive Industries Accounting 1 (Summer 1982): 47-57.

Stanger, Abraham M. "Oil and Gas Accounting: Who Makes the Rules?" Corporation Law Review 2 (Spring 1979): 154-158.

Stolle, Carlton, D. "The Coming of Current Replacement Cost in the United States." Oil & Gas Tax Quarterly 25 (September 1976): 73-98.

Sunder, Shyam. "Properties of Accounting Numbers Under Full Costing and Successful Efforts Costing in the Petroleum Industry." Accounting Review 51 (January 1976): 1-18.

Trueman, J. W. H. "Oil Company Accounts: Not So Comparable?" Accounting and Business Research (Eng.) 5 (Spring 1975): 127-132.

"U.S. Reporting Rules Obscuring Important Issues." Accountancy (Eng.) 89 (July 1978): 12.

Warne, Richard A. "Controller Looks at Regulation." Management Accounting 63 (June 1982): 14-20.

Weiser, Herbert J. "A Movement Toward Fair and Meaningful Reporting." Management Accounting 58 (February 1977): 13-16.

Welsch, Glenn A., and Deakin, Edward B. "Usefulness, Objectivity, or Both? Trade-Off in Accounting for the Oil and Gas Industry." Oil & Gas Tax Quarterly 26 (December 1977): 203-223.

Whittred, G. P. "Accounting for the Extractive Industries: Use or Abuse of the Matching Principle?" Abacus (Australia) 14 (December 1978): 154-159.

"Who Will Set CPA Rules." Dun's Review 108 (September 1976): 62-65.

Wick, M. S. "The New York Stock Exchange and Financial Disclosure." The Accounting Forum 43 (May-December 1973): 9-16.

Wilkerson, Curtis A., and McEntee, Matt. "A Review of FASB Interpretation No. 36." Journal of Extractive Industries Accounting 1 (Fall/Winter 1982): 47-55.

Wright, Charlotte. "Accounting for Reclamation Costs of Oil and Gas Operations." Journal of Extractive Industries Accounting 1 (Spring 1982): 77-105.

"You Can't Legislate Accounting Principles." Forbes 122 (October 2, 1978): 90, 93, 95.

DISCLOSURES AND RESERVE RECOGNITION ACCOUNTING

Adkerson, Richard C. "Can Reserve Recognition Accounting Work?" Journal of Accountancy 146 (September 1978): 72-81.

Alford, R. Mark; Benjamin, James J.; and Strawser, Robert H. "Disclosures Concerning Oil and Gas Producing Activities and the FASB." Oil & Gas Tax Quarterly 30 (June 1982): 685-702.

Avard, Stephen L. "Disclosures of Reserve Quantities, Reserve Values, and Performance Measures--Views of Financial Analysts." Journal of Extractive Industries Accounting 1 (Spring 1982): 71-75.

Avard, Stephen L. "Financial Analysts' Evaluation of Proposed Disclosure Rules for Oil and Gas Producing Companies." Journal of Extractive Industries Accounting 1 (Fall/Winter 1982): 125-131.

Barton, M. Frank, Jr.; Calhoun, Charles H.; and Carper, William B. "Some Preliminary Findings Concerning Reserve Recognition Accounting in the Oil and Gas Industry." Oil & Gas Tax Quarterly 28 (December 1979): 216-227.

Bybee, W. Cecil. "Valuation Accounting for Oil and Gas." Viewpoint (1978): 7-9.

Connor, Joseph E. "Discovery Value--The Oil Industry's Untried Method." Journal of Accountancy 139 (May 1975): 54-63.

Connor, Joseph E. "Reserve Recognition Accounting: Fact or Fiction?" Journal of Accountancy 148 (September 1979): 92-99.

Cooper, Kerry; Flory, Steven M.; Grossman, Steven D.; and Groth, John C. "Reserve Recognition Accounting: A Proposed Disclosure Framework." Journal of Accountancy 148 (September 1979): 82-91.

Davies, Barry J. "Practical Problems Preclude Meaningful Disclosure." Chartered Accountant in Australia 50 (April 1980): 19-20, 23-25.

Deakin, Edward B., and Detrick, James W. "An Evaluation of RRA and other Supplemental Oil and Gas Disclosures by Financial Anaylsts." Journal of Extractive Industries Accounting 1 (Spring 1982): 63-70.

Dugas, Kenneth L., and Posey, Clyde L. "Reserve Recognition in Accounting." Mid-South Business Journal 1 (October 1981): 19-23, 28.

Eddey, Peter H., and Sullivan, Graham G. "Disclosure of Reserves by Oil and Gas Companies." Chartered Accountant in Australia 52 (November 1981): 15-19.

Flory, Steven M., and Brown, Richard. "Delayed Reserve Valuation for Oil and Gas Companies." Oil & Gas Tax Quarterly 28 (March 1980): 341-353.

Fraser, Lyn M. "RRA: A Look Behind the Theory to the Numbers." Journal of Accountancy 148 (September 1979): 106-111.

Greene, Richard. "Buried Treasure--Reserve Recognition System." Forbes 126 (September 29, 1980): 109, 113.

Grier, J. D., and Miglicco, G. L. "Reserve Recognition Accounting: How It Works." Oil & Gas Tax Quarterly 29 (September 1980): 120-138.

Grossman, Steven D.; Shugart, Gary L.; and Strawser, Robert H. "A Study on Replacement Cost Disclosure in Petroleum Industry." Oil & Gas Tax Quarterly 26 (March 1978): 332-345.

Hall, James D. "A Closer Look at RRA." Oil & Gas Journal (January 7, 1980): 73-74.

Heazlewood, C. T. "Mineral Reserves--Should Disclosure be Mandatory: A Case for Following the U.S. Example." Chartered Accountant in Australia 50 (April 1980) 14-15, 17-18.

Kelley, Robert. "Accounting for the Effects of Reserve Changes on Interim Results." Journal of Extractive Industries Accounting 1 (Fall/Winter 1982): 11-19.

Keplinger, H. F. "Reserve Calculation Review Shows Why Engineering Estimates Differ." Oil & Gas Journal (January 17, 1977): 60-70.

Keplinger, H. F. "Toward Understanding Oil and Gas Reserves." Public Utilities Fortnightly 99 (February 3, 1977): 18-23.

King, Barry G., and Brock, Horace R. "Financial Disclosures for Oil and Gas Producers." Oil & Gas Tax Quarterly 29 (June 1981): 759-780.

King, Barry G. "Oil and Gas Disclosure: Some Empirical Results." Journal of Extractive Industries Accounting 1 (Spring 1982): 107-127.

Koester, Robert J., and Chisholm, Sam W. "Reserve Recognition Accounting: Lodestar for the Future?" Oil & Gas Tax Quarterly 28 (September 1979): 86-98.

Koester, Robert J.; Munter, Paul; and Ratcliffe, Thomas A. "Current Status of Reserve Recognition Accounting." Oil & Gas Tax Quarterly 30 (December 1981): 341-354.

May, Alan, Jr., and O'Neill, Donald B. "Internal Audit of Oil and Gas Reserves." Journal of Extractive Industries Accounting 1 (Fall/Winter 1982): 37-45.

Moore, Cecil H., and Duncan, Alfred G., Jr. "Rah Rah--Is Reserve Recognition Accounting a Change for the Better?" World (Autumn 1979): 30-36.

Most, Kenneth S. "A New Method of Accounting for Oil and Gas Producers." Management Accounting, (May 1979): 53-58.

Munter, Paul; Ratcliffe, Thomas A.; and Koester, Robert J. "Reporting on Supplementary Energy Reserve Information." Oil & Gas Tax Quarterly 31 (September 1982): 150-177.

Nemeth, L. K., and Lamme, C. L. "RRA--New Requirements Are Costly, Demanding." World Oil 188 (January 1979): 93-94.

Phillips, Joseph I., and Waldron, Robin S. "Reserve Recognition Accounting--Resource and Energy Truths." Certified Accountant (Eng.) 73 (August 1981): 195-198, 200-201, 254.

Seidler, Lee J. "Gulf Oil's RRA Example: Distorted." Accounting Issues (January 18, 1979): 4-17.

Siegel, Joel G., and Geller, Brent. "Accounting for Oil and Gas Reserves: Analysis and Recommendations." Ohio CPA Journal 39 (Spring 1980): 79-80.

Society of Petroleum Engineers. "Exposure Draft: Standards Pertaining to the Estimating and Auditing of Oil and Gas Reserve Information." Journal of Petroleum Technology (July 1979): 852-860.

Society of Petroleum Engineers. "Standards Pertaining to the Estimating and Auditing of Oil and Gas Reserve Information." Journal of Petroleum Technology 31 (December 1979): 1557-1565.

Stolle, Carlton, D. "SEC Reporting: A Proposal Affecting Reporting of Oil and Gas Reserves." Oil & Gas Tax Quarterly 24 (June 1976): 533-537.

Stuart, Alexander. "Does Anybody Really Need R.R.A.?" Fortune 99 (June 18, 1979): 70-82.

"Survey of Oil and Gas Disclosures." CPA Journal 49 (December 1979): 7-8.

Walendowski, George, and Diamond, Michael A. "RRA--Will It Work?" Management Accounting 61 (March 1980): 21-25.

Wallace, J. Rodney. "Oil and Gas Accounting--Have We Found the Missing Assets?" Accountant's Magazine (Scot.) 86 (May 1982): 159-160, 180.

Wallace, J. Rodney. "Oil and Gas Accounting--the Missing Assets?" Accountant's Magazine (Scot.) 85 (December 1981): 416-419.

Walther, Larry, and Evans, Mary Kay. "A Study of Revisions of Previous Estimates of Proved Oil and Gas Reserve Quantity Information." Journal of Extractive Industries Accounting 1 (Fall/Winter 1982): 99-108.

Wright, Charlotte. "A Study of Effects of News Reports on Stock Prices: Implications for Reserve Quantity Disclosures." Journal of Extractive Industries Accounting 1 (Fall/Winter 1982): 109-115.

DRILLING FUNDS, PROGRAMS, JOINT VENTURES, AND PARTNERSHIPS

Baggett, Monte R.; Dole, Richard D.; and Short, Jack E. "Accounting and Auditing for Drilling Funds." CPA Journal 51 (September 1981): 27-28, 32-34, 36-37. Excerpted from Anatomy of a Drilling Fund, Coopers & Lybrand, 1980.

Berner, Arthur S., and Scoggins, Sue. "Oil and Gas Drilling Programs--Structure and Regulation." George Washington Law Review 41 (March 1973): 471-504.

Guest, J. Ike. "Financial Reporting by Drilling Funds." Journal of Extractive Industries Accounting 1 (Fall/Winter 1982): 21-36.

Johnson, Steven. "An Accounting System for Joint Ventures." Management Accounting 54 (April 1973): 37-38.

McDonald, Malcolm S. "Joint Venture Accounting in the North Sea." Accountant's Magazine (Scot.) 84 (July 1980): 277-279.

Patrick, Don A. "Auditing Joint Interest Operations in Periods of Shortages or Oversupply of Materials." Journal of Extractive Industries Accounting 1 (Fall/Winter 1982): 57-67.

Valente, Renato. "Oil Royalties and Oil Royalty Funds." Financial Planning Today 4 (November 1980): 313-321.

ECONOMICS, FINANCE, AND EVALUATION

Akins, James E. "New Directions in Energy Policy." *Financial Analysts Journal* 29 (January/February 1973): 29-32, 61-63, 92.

Beck, Robert J., and Tippee, Bob. "Big Oil Firm Profits Mask Dilemma of U.S. Reserves." *Oil & Gas Journal* (May 26, 1980): 31-36.

Carlton, Alan. "How to Evaluate a Producing Property for Purchase." *Oil & Gas Journal* (September 15, 1980): 105.

Castle, Grover R. "Lending to the Small Independent Oil and Gas Operator." *Journal of Commercial Bank Lending* 61 (February 1979): 9-24.

Collins, Daniel W.; Rozeff, Michael S.; and Dhaliwal, Dan S. "Economic Determinants of the Market Reaction to Proposed Mandatory Accounting Changes in the Oil and Gas Industry: A Cross-Sectional Analysis." *Journal of Accounting & Economics* (Netherlands) 3 (March 1981): 37-71.

Copp, E. Anthony. "Technical Change and Petroleum Drilling Costs." *Land Economics* 50 (May 1974): 120-126.

Copp, E. Anthony. "Technical Change and Petroleum Drilling Costs: Reply." *Land Economics* 52 (February 1976): 109-116.

Cowen Scott S., and Hoffer, Jeffrey A. "Usefulness of Financial Ratios in a Single Industry." *Journal of Business Research* 10 (March 1982): 103-118.

Cox, James C., and Wright, Arthur W. "Determinants of Investment in Petroleum Reserves and Their Implications for Public Policy." *American Economic Review* 66 (March 1976): 153-167.

"Credit System Speeds Customer Service." *Infosystems* 25 (April 1978): 104, 106.

Cudaback, David. "Energy Financing Battle Heats Up." *Institutional Investor* 14 (July 1980): 89-91, 93, 95, 98.

Dhaliwal, Dan S. "The Effect of the Firm's Capital Structure on the Choice of Accounting Methods." *Accounting Review* 55 (January 1980): 78-84.

Farnsworth, Richard L., and Norgaard, Richard B. "Technical Change and Petroleum Drilling Costs: Comment." Land Economics 52 (February 1976): 103-108.

Frair, Les, and Devine, Mike. "Economic Optimization of Offshore Petroleum Development." Management Science 21 (August 1975): 1370-1379.

Jackson, H. Noel. "U.S. Investments in Middle East Petroleum." California Management Review 17 (Summer 1975): 22-27.

Kuller, Robert G., and Cummings, Ronald G. "Economic Model of Production and Investment for Petroleum Reservoirs." American Economic Review 64 (March 1974): 66-79.

Lappen, Alyssa A. "Appraised Value: The Stuff of Dreams." Forbes 125 (May 12, 1980): 145-146, 151-152, 154.

Levy, Walter J. "We Need a New Financing Instrument." Institutional Investor 14 (July 1980): 119-120, 122, 124.

Mao, James C. T. "Project Financing: Funding the Future." Financial Executive 50 (August 1982): 23-24, 26, 28.

Mead, David E. "Effect of Vertical Integration on Risk in the Petroleum Industry." Quarterly Review of Economics & Business 18 (Spring 1978): 83-90.

Morthland, Rex J. "Banker and the Energy Crisis." Journal of Commercial Bank Lending 56 (June 1974): 13-21.

Moyer, Reed. "Competitive Impact of the Emerging Energy Company." California Management Review 16 (Summer 1974): 81-92.

Nelms, Terry L. "Oil & Gas Properties: Trust Asset or Liability?" Trusts & Estate 115 (August 1976): 520-523.

Noonan, Edward R. "1973 Accounting Principles--Evolution or Revolution." Financial Forum 11 (April 1973): 1-5. (Reprinted from The Journal of Commercial Bank Lending, January 1973).

Norris, John W. "Lending to the Energy Industry." Journal of Commercial Bank Lending 64 (May 1982): 31-40.

Perille, P. James, and Saathoff, Frederick J. "Why Not Project Financing?" Management Accounting 60 (October 1978): 13-16, 22.

Phillips, Paul D.; Groth, John C.; and Richards, R. Malcolm. "Financing the Alaskan Project: The Experience at Sohio." Financial Management 8 (Autumn 1979): 7-16.

Reid, Samuel Richardson. "Petroleum Mergers, Multinational Investments, Refining Capacity and Performance in the Energy Crisis." Financial Management 2 (Winter 1973): 50-56.

Richards, Paul H. "Project Financing in the North Sea." International Accountant (Eng.) 46 (No. 4, 1976): 2-4.

Siegel, J. G. and Savage, L. "Industry Characteristics and the Quality of Earnings." National Public Accountant 24 (July 1979): 11-15.

Smallwood, J. William. "Investment Policy for the Energy Crisis of the 1970's." Trusts and Estates 112 (February 1973): 118-121.

Smith, Rodney T. "An Economic Analysis of Income Growth by U.S. Oil Firms: The Roles of U.S. Oil Regulation and OPEC." Journal of Business 55 (October 1982): 427-478.

Sultan, Ralph G. M. "Dimensions of the Investment Boom." Canadian Business Review 1 (Summer 1974): 18-21.

Swinth, Robert L.; Gaumnitz, Jack E.; and Rodriguez, Carlos. "Decision Making Processes: Using Discriminant Nets for Security Selection." Decision Sciences 6 (July 1975): 439-448.

Townsend, Paul S. "Appraisal of Petroleum Storage and Distribution Centers." Appraisal Journal 46 (April 1978): 165-175.

Walther, Larry M. "Commodity Futures: What the Accountant Should Know." Journal of Accountancy 153 (March 1981): 68-70, 73-74, 76, 78-80, 82.

Wilson, John W. "Adam Smith Abandoned: Big Oil Is Big Coal Is Big Natural Gas." Business and Society Review/Innovation (Spring 1974): 65-72.

Wynant, Larry. "Project Financing for Extractive Ventures." Management Accounting 60 (October 1978): 29-35.

PLANNING, CONTROL, AND BUDGETING

Babayev, Djangir A. "Mathematical Models for Optimal Timing of Drilling on Multilayer Oil and Gasfields." Management Science 21 (August 1975): 1361-1369.

Bell, S. Kenneth. "A Survey of Planning Practices in the Petroleum Industry." Managerial Planning 23 (September/October 1974): 29-32.

Blaker, N. "Management Accounting in the Gas Industry." Public Finance & Accountancy (Eng.) 2 (April 1975): 119-123.

Cozzolino, John M. "Controlling Risk in Capital Budgeting: A Practical Use of Utility Theory for Measurement and Control of Petroleum Exploration Risk." Engineering Economist 25 (Spring 1980): 161-186.

Dallenbach, Hans G. "Model of A Multi-Product Two-Stage Inventory System with Limited Intermediate Bulk Storage Capacity." Management Science 23 (August 1977): 1314-1320.

Deakin, Edward B. "Auditor Selection, Organization Control, Adverse Events, and the Selection of Accounting Method for Oil Exploration." Quarterly Review of Economics and Business, 20 (Autumn 1980): 77-85.

Devine, Michael D. "Model for Minimizing the Cost of Drilling Dual Completion Oil Wells." Management Science 20 (December 1973, part I): 532-535.

Durrer, E. J. and Seater, G. E. "Optimization of Petroleum and Natural Gas Production--A Survey." Management Science 24 (September 1977): 35-43.

Feller, Robert E. "Accounting for Joint Products in the Petroleum Industry." Management Accounting 59 (September 1977): 41-44, 48.

Gartenberg, Morris. "How Dow Accounts for Its Energy Use." Management Accounting 61 (March 1980): 10-12.

Ihlanfeldt, W. J. "Internal Control--Shell Oil Company's Approach." Financial Executive 47 (July 1979): 42-46, 48.

Krasts, Aivars. "Coping with Inflation in Capital Budgeting and Economic Performance Evaluation." Managerial Planning 31 (July/August 1982): 16-20.

Krasts, Aivars. "Performance Evaluation at Conoco." Planning Review 9 (November 1981): 30-33.

Lilien, Gary L. "Note on Offshore Oil Field Development Problems and Suggested Solutions." Management Science 20 (December 1973, part I): 536-539.

May, Alan, Jr., and O'Neill, Donald B. "Internal Audit of Oil and Gas Reserves." Journal of Extractive Industries Accounting 1 (Fall/Winter 1982): 37-45.

Neumann, Bruce R., and Landagora, Dustin C. "Measuring Divisional Performance for an Oil Company." Management Accounting 63 (March 1982): 41-45.

Rubin, Harvey W., and Daniel, Ronald W. "Managing Fuel Oil Distribution Risks." Risk Management 23 (October 1976): 40-42, 44, 48.

Warne, Richard A. "Controller Looks at Regulation." Management Accounting 63 (June 1982): 14-20.

INTERNATIONAL

Amernic, Joel H. "Accounting Practices in the Canadian Petroleum Industry." CA Magazine (Canada) 112 (March 1979): 34-38.

Bailey, Richard. "Must We Share North Sea Oil?" Accountant (Eng.) 179 (October 26, 1978): 542-543.

Briers, Roger. "Analyzing Business Opportunities in North Sea Oil." Long Range Planning 9 (February 1976): 53-58.

Brown, Andrew. "The Fourth Dimension." Canadian Business 55 (June 1982): 30-31.

Chang, Lucien E., and Most, Kenneth S. "International Accounting Standards: The Case of European Oil Companies." International Journal of Accounting Education and Research 12 (Fall 1976): 27-43.

Davison, A. G. and Lourens, R. M. C. "Compliance with DSR (points from published accounts)." Chartered Accountant in Australia 48 (May 1978): 33-36.

Glendinning, Robert. "Accounting in the Nationalized Industries." Management Accounting (Eng.) 60 (April 1982): 32-33.

Heazlewood, C. T. "DSR, October 1976, Doomed to Failure from the Start." Chartered Accountant in Australia 49 (July 1978): 22-24.

Heazlewood, C. T. "Mineral Reserves--Should Disclosure be Mandatory: A Case for Following the U.S. Example." Chartered Accountant in Australia 50 (April 1980) 14-15, 17-18.

Holland, F. A.; Wilkinson, J. K.; and Watson, F. A. "No Salvation in North Sea Oil." Accounting (Eng.) 86 (April 1975): 40-42.

McDonald, Malcolm S. "Joint Venture Accounting in the North Sea." Accountant's Magazine (Scot.) 84 (July 1980): 277-279.

Morley, Michael F. "Whose Offshore Oil?" CA Magazine (Canada) 111 (July 1978): 15-16.

Most, Kenneth S. "A Survey of the 1973 Annual Reports of Eight European Oil Companies." Accounting Forum 46 (May 1976): 7-16.

Mowat, Malcolm. "Accounting for Costs in the UK Oil Industry." Accountant's Magazine (Scot.) 84 (June 1980): 229-231.

Murray, Joe. "Opportunities Offshore for Irish Industry." Management (Irish) 23 (June 1976): 53-54, 57.

Nethercott, Leslie, J. "Discovery Value Accounting--A Viable Alternative?" Australian Accountant 47 (May 1977): 228-231.

Nethercott, Leslie J. "Oil--Is Discovery Value Accounting the Answer?" Accountancy (Eng.) 87 (July 1976): 28, 30, 32.

Potter, C. C. "Who Sets Standards?" Australian Accountant 50 (January/February 1980): 26-28.

Richards, Paul H. "Project Financing in the North Sea." International Accountant (Eng.) 46 (No. 4, 1976): 2-4.

Robinson, Chris. "How the CICA Should Respond to Canada's Growth." CA Magazine (Canada) 115 (January 1982): 30-33, 35.

Rolland, M. J. "North Sea Oil and the Scottish Economy." Accountant's Magazine (Scot.) 77 (February 1973): 92-95.

Sultan, Ralph G. M. "Dimensions of the Investment Boom." Canadian Business Review 1 (Summer 1974): 18-21.

ELECTRONIC DATA PROCESSING

Perry, William E. "Trends in EDP Auditing." EDPACS 4 (December 1976): 1-8.

Sarson, Trish, and Gane, Chris. "Structured Methodology: What Have We Learned?" Computerworld 14 (September 17, 1980): 52-57.

Willis, David W. "Oil and Gas Accounting Systems: Going On-Line." Journal of Extractive Industries Accounting 1 (Summer 1982): 73-77.

AUDITING

Baggett, Monte R.; Dole, Richard D.; and Short, Jack E. "Accounting and Auditing for Drilling Funds." CPA Journal 51 (September 1981): 27-28, 32-34, 36-37. Excerpted from Anatomy of a Drilling Fund, Coopers & Lybrand, 1980.

Deakin, Edward B. "Auditor Selection, Organization Control, Adverse Events, and the Selection of Accounting Method for Oil Exploration." Quarterly Review of Economics and Business, 20 (Autumn 1980): 77-85.

DeAngelo, Linda Elizabeth. "Mandated Successful Efforts and Auditor Choice." Journal of Accounting and Economics 4 (December 1982): 171-203.

Doblin, Clair P. "Multinational Corporations and Government Auditing." International Journal of Government Auditing 2 (January 1975): 9-12.

May, Alan, Jr., and O'Neill, Donald B. "Internal Audit of Oil and Gas Reserves." Journal of Extractive Industries Accounting 1 (Fall/Winter 1982): 37-45.

Moore, Cecil H., and Button, Damon. "Auditing Implications of DOE Controls." Journal of Accountancy 148 (August 1979): 78-83

Patrick, Don A. "Auditing Joint Interest Operations in Periods of Shortages or Oversupply of Materials." Journal of Extractive Industries Accounting 1 (Fall/Winter 1982): 57-67.

Perry, William E. "Trends in EDP Auditing." EDPACS 4 (December 1976): 1-8.

Reynolds, Allan. "Expanded Scope Audits in the Department of the Interior." Government Accountants Journal 26 (Fall 1977): 14-17.

Weirich, Thomas R., and Grant, Edward B. "Broadening Financial Reporting: The Auditor's Involvement." Journal of Accountancy 152 (November 1981): 104-118.

GOVERNMENT REGULATION AND PRICE CONTROL

Michigan Law Review. "Refunding Overcharges Under the Emergency Petroleum Allocation Act: The Evolution of a Compensatory Obligation." Michigan Law Review 79 (June 1981): 1454-1476.

Moore, Cecil H., and Button, Damon. "Auditing Implications of DOE Controls." Journal of Accountancy 148 (August 1979): 78-83.

Munter, Paul; Koester, Robert J.; and Ratcliffe, Thomas A. "Streamlining Department of Energy Financial Reporting." Oil & Gas Tax Quarterly 30 (March 1982): 544-557.

Nash, Susan E. "Collecting Overcharges from the Oil Companies: The Department of Energy's Restitutionary Obligation." Stanford Law Review 32 (May 1980): 1039-1059.

REFINING

Al-Saffar, Hadi R. "Management Accounting in an Oil Refinery." Management Accounting (Eng.) 60 (January 1982): 22-25.

El-Badawi, Mohamed H. "Effect of Product Costing on Refinery Income Determination." Oil & Gas Tax Quarterly 30 (March 1982): 520-543.

PIPELINING AND TRANSMISSION

Elting, William W., and Gaynor, Joseph P. "Diversification and Stock Market Performance in the Natural Gas Industry." Public Utilities Fortnightly 100 (August 18, 1977): 18-24.

Jones, G. A., and Wilson, J. G. "Optimal Scheduling of Jobs on a Transmission Network." Management Science 25 (January 1979): 98-104.

Rasmussen, William G. "Alabama Gas Corp. Estimates Unbilled Energy." Management Accounting 62 (February 1981): 23-27.

Twizell, D. "Customer Accounting in the Gas Industry." Public Finance and Accounting (Eng.) 6 (February 1979): 32-36.

INTEREST CAPITALIZATION

Arcady, Alex T., and Klein, John R. "Capitalizing Interest for Oil and Gas Companies: The FASB's Answer." Oil & Gas Tax Quarterly 29 (September 1980): 109-119.

Bowen, Robert M.; Noreen, Eric W.; and Lacey, John M. "Determinants of the Corporate Decision to Capitalize Interest." Journal of Accounting & Economics (Netherlands) 3 (August 1981): 151-179.

Martin, D. Edward. "Something New in Accounting: Capitalizing Interest." Financial Executive 48 (May 1980): 24-28.

Munter, Paul, and Ratcliffe, Thomas A. "Accounting for Interest Costs by Oil And Gas Producers." Oil & Gas Tax Quarterly 31 (December 1982): 452-461.

OIL AND GAS DISSERTATIONS 1973-1982

Bell, Timothy Barnes. "Market Reaction to Reserve Recognition Accounting." Ph.D. dissertation, Oklahoma State University, 1981.

Burnaby, Priscilla Ann. "The Information Content of Reserve Recognition Accounting." Ph.D. dissertation, Texas A & M University, 1982.

DuBois, Donald Ayers. "Full Cost Accounting for Exploration and Development Costs of Petroleum Companies--Its Effect on Stock Market Prices." Ph.D. dissertation, University of Missouri.

Elmadfai, Abdulfattah Omar Dia. "An Integrated Theory of Accounting Using Capital/Expense Decision in the Petroleum Industry as a Special Application." Ph.D. dissertation, Louisiana State University, 1976.

Eskew, Robert Kendrick II. "An Empirical Examination of the Interaction Between Accounting Alternatives and Share Price in the Extractive Petroleum Industry." Ph.D. dissertation, Purdue University, 1973.

Etebari-Khoransgani, Ahmad. "An Assessment of the Stock Market Effects of Proposed Changes in the Oil and Gas Industry." Ph.D. dissertation, North Texas State University, 1979.

Haskin, Daniel Lee. "An Investigation of the Usefulness of Reserve Recognition Accounting Data." D.B.A. dissertation, Texas Tech University, 1982.

Huang, Jiunn-Chang. "An Empirical Investigation into the Information Content of the Required Disclosure of Oil and Gas Reserve Values." Ph.D. dissertation, North Texas State University, 1980.

Lys, Thomas. "Selection of Accounting Procedures and Implications of Changes in Generally Accepted Accounting Principles: A Case Study Using Oil and Gas Accounting." Ph.D. dissertation, University of Rochester, 1982.

Mannino, Ronald Carl. "The Impact of the Use of the Full Cost Method in the Petroleum Industry on the Market Price of Securities." Ph.D. dissertation, University of Colorado at Boulder, 1977.

Mohammadzadeh-Novin, Adel. "An Empirical Investigation of the Information Content of the Disclosure on the Value of Oil and Gas Reserves." Ph.D. dissertation, University of Georgia, 1982.

Naggar, Ali K. "An Empirical Study of Evaluating 'Successful Efforts' and 'Full Cost' Methods of Accounting for Finding Cost of Oil and Gas Industry from the Investor's Point of View." Ph.D. dissertation, University of Oklahoma, 1975.

Oien, M. Burton. "A Comparison of Cost and Market Based Accounting Models for a Major Petroleum Company." Ph.D. dissertation, University of Oklahoma, 1976.

Ormsby, Susan Young. "By-Product Accounting in the Extractive Industries: A History and Survey of Current Practice." Ph.D. dissertation, University of Arkansas, 1982.

Peyvandi, Ali Agha. "Accounting and Reporting for Investment in Joint Venture Arrangements." Ph.D. dissertation, University of Missouri--Columbia, 1980.

Schmeltz, William Frederick. "Accounting and Management Control Practices in Petroleum Refining." Ph.D. dissertation, Western Reserve University, 1966.

Smith, Abbie. "An Empirical Investigation of the Information Effects of a Change in the Financial Reporting Standards for Oil and Gas Producers: The Proposed Elimination and Subsequent Retention of Full Cost Accounting." Ph.D. dissertation, Cornell University, 1981.

Tarabzune, Muhiadin Rashad. "An Analysis of the Effect of Capitalizing Exploration and Development Costs in the Petroleum Industry--with an Emphasis on Possible Economic Consequences in Saudi Arabia." Ph.D. dissertation, University of Arkansas, 1975.

Tondkar, Rasoul H. "An Empirical Investigation of the Potential Use of Data Required by FASB Statement No. 33 by Financial Analysts in the Dallas/Ft. Worth and Houston Regions." Ph.D. dissertation, North Texas State University, 1980.

Wright, Charlotte Jean. "An Assessment of the Effect of News Anouncements on Stock Prices of Oil and Gas Producing Companies." Ph.D. dissertation, North Texas State University, 1982.

1973-1982

Mining

MINING
1973-1982

GENERAL

Arthur Andersen and Company. Coal: Industry Reference Binder. Chicago, 1980.

Chakraborty, S. K. "Accounting for Overburden Removal in Opencast Mines." Management Accountant (India) 11 (February 2, 1976): 118-122.

Cleaver, George H. "Nonferrous Metals," in Levine, Sumner N. Financial Analyst's Handbook. Homewood, Ill., 1975, v. 2, 524-543.

Clemison, John. "Accounting for Basic Industry: Coal Mining by Numbers." Accountant (Eng.) 184 (May 28, 1981): 590-592.

Cook, James. "It's Ethical, But Is It Fair?" Forbes 118 (September 15, 1976): 107-108.

Coopers & Lybrand. Financial Reporting in the Mining Industry: An International Survey. New York, 1979.

Davison, A. G. "Mining Companies' Financial Statements--Of What Use?" Chartered Accountant in Australia 50 (August 1979): 22-23, 25-26.

Ezejelue, A. C., and Ofodile, C. O. "Accounting for Extractive Operations--The Case of the Nigerian Coal Corporation." Chartered Accountant (India) 29 (April 1981): 765-769, 779.

Fagerberg, Dixon, "Metal Mining Companies," in Encyclopedia of Accounting Systems. Rev. ed. Englewood Cliffs, N.J., 1976, 1269-1310.

Financial Accounting Standards Board. Mining Task Group. Interim Report on FASB Exposure Drafts: Financial Reporting and Changing Prices and Constant Dollar Accounting. Stamford. Conn., May 31, 1979.

French, E. "Physical Capital Maintenance in Income Measurement--An Exemplification." Accounting and Business Research 6 (Winter 1976): 36-44.

Hedge, P. G. "Costing in Mining Industry." Management Accountant (India) 15 (August 1980): 327-330, 335.

Janson, Ernest C. "Mining: Comment on Interim Task Group Report," in Proceedings, Conference on Financial Reporting and Changing Prices. Stamford, Conn., 1979, 87-91.

Janson, Ernest C.; Krup, Stephen C.; and Wright, David T. Financial Reporting and Tax Practices in Nonferrous Mining. 8th ed. New York: Coopers & Lybrand, 1979.

Janson, Ernest C.; Krup, Stephen C.; and Wright, David T. Financial Reporting and Tax Practices in Nonferrous Mining. 9th ed. New York: Coopers & Lybrand, 1980.

Janson, Ernest C.; Krup, Stephen C.; and Wright, David T. Financial Reporting and Tax Practices in Nonferrous Mining. 10th ed. New York: Coopers & Lybrand, 1981.

Janson, Ernest C.; MacLean, John A.; and Wright, David T. Financial Reporting and Tax Practices in Nonferrous Mining. 3rd ed. New York: Coopers & Lybrand, 1974.

Janson, Ernest C.; MacLean, John A.; and Wright, David T. Financial Reporting and Tax Practices in Nonferrous Mining. 4th ed. New York: Coopers & Lybrand, 1975.

Janson, Ernest C.; MacLean, John A.; and Wright, David T. Financial Reporting and Tax Practices in Nonferrous Mining. 5th ed. New York: Coopers & Lybrand, 1976.

Janson, Ernest C.; MacLean, John A.; and Wright, David T. Financial Reporting and Tax Practices in Nonferrous Mining. 6th ed. New York: Coopers & Lybrand, 1977.

Janson, Ernest C.; MacLean, John A.; and Wright, David T. Financial Reporting and Tax Practices in Nonferrous Mining. 7th ed. New York: Coopers & Lybrand, 1978.

Krilich, James H. "Look at the Copper Industry." Management Accounting 57 (March 1976): 37-38.

Lourens, Roy. "Principles, Profits--And Problems." South African Chartered Accountant 11 (May 1975): 162-165.

Lucas, J. Richard; Shaw, C. Timothy; and Lightner, M. William. "Coal." Mahon's Industry Guides for Accountants and Auditors No. 14. Boston, 1980.

Matheny, James L. "Is GAAP Ready for the Coal Boom?" Pennsylvania CPA Spokesman 49 (April 1979): 10-14.

Mayorshi, Frank B., and Prickitt, Kenneth L. "We Dig Coal." Virginia Accountant 27 (September 1975): 15-22.

McCarl, Henry N.; MacLean, John A.; and Wright, David T. "Non-Metallic Minerals." Mahon's Industry Guides for Accountants and Auditors No. 16. Boston, 1980.

McNally, Edward T. "Cost of Coal--A Conjecture." Conference Board Record 12 (January 1975): 57-59.

Muir, W. L. Reclamation of Surface Mined Land. England: Muir Coal Industry Information Service, 1979.

Nethercott, Leslie J. "Depletion in the Mining Industry." Australian Accountant 43 (February 1973): 20-22.

Nethercott, L. J. "Full Cost Accounting in the Mining Industry." Australian Accountant 45 (November 1975): 572-577.

Peat, Marwick, Mitchell & Company. Mining Survey: Review of 1975 Annual Reports to Stockholders. New York, 1976.

Peat, Marwick, Mitchell & Company. Principles and Presentation: Mining. New York, 1977.

Peat, Marwick, Mitchell & Company. Principles and Presentation: Mining. New York, 1978.

Peat, Marwick, Mitchell & Company. Principles and Presentation: Mining. 1978 ed. New York, 1979.

Peat, Marwick, Mitchell & Company. Principles and Presentation: Mining. New York, 1981.

Randall, Alan, et al. "Reclaiming Coal Surface Mines in Central Appalachia: A Case Study of the Benefits and Costs." Land Economics 54 (November 1978): 472-489.

Seth, Shirish B., and Woo, John C. H. "The Impact of Replacement Cost Accounting on Financial Performance." Financial Analysts Journal 34 (March/April 1978): 48-54.

Swinney, R. S. Results of the 1980 Survey of Accounting Practices in the Coal Industry. Washington: National Coal Association, 1980.

Straws, Simon D.; Janson, Ernest C.; and Wright, David T. "Non-Ferrous Metals." Mahon's Industry Guides for Accountant's and Auditors No. 15. Boston, 1980.

Walker, Michael A. "Form S-18 Amended for Mining Companies and Form S-3 Rescinded." CPA Journal 51 (September 1981): 83.

Woods, E. M., and Brooking, D. L. "In Our Opinion: Monthly Comment on Annual Reports." South African Chartered Accountant 17 (May 1981): 261-262.

Zimmermann, Martin B. "Modeling Depletion in a Mineral Industry: The Case of Coal." Bell Journal of Economics 8 (Spring 1977): 41-65.

News Releases and Short Comments

"Metal/Mining Pension Plans Analyzed." Pension & Welfare News 10 (February 1974): 54, 57-58.

"SEC's SAB No. 18 Describes Mining Operation Disclosure." Journal of Accountancy 144 (December 1977): 24.

Beresford, Dennis R., and Neary, Robert D. "Guides Issued for Disclosing Mining Replacement Costs." Financial Executive 46 (January 1978): 17.

"Main Issue is the Valuation of Fixed Assets--National Coal Board." Public Finance and Accountancy (Eng.) 6 (September 1979): 39-40.

"Mine of Information." Accountant (Eng.) 182 (April 24, 1980): 624-625.

Beresford, Dennis R., and Neary, Robert D. "SEC Registration and Reporting by Mining Companies." Financial Executive 48 (December 1980): 9-10.

Beresford, Dennis R., and Neary, Robert D. "SEC Postpones Proposed Mining Disclosures." Financial Executive 49 (June 1981): 9.

LEASES, ROYALTY CLAUSES, FARMOUTS, AND OTHER AGREEMENTS

Adams, Mark K. "Minimum Work Clauses in Mining Leases," in Rocky Mountain Mineral Law Institute, 21st, 1975 Proceedings. New York: Matthew Bender & Company, 1975, 535-564.

Alfers, Stephen D. "The Uranium Royalty Today," in Rocky Mountain Mineral Law Institute, 25th, 1979 Proccedings. New York: Matthew Bender & Company, 1979, 19-1--19-84.

Carney, Edmund M. "Pricing Provisions in Coal Contracts," in Rocky Mountain Mineral Law Institute, 24th, 1978 Proceedings. New York: Matthew Bender & Company, 1978, 197-232.

Clark, Graham M. "Economic Considerations for the Layman/Landman in Negotiating Mining Leases," in Rocky Mountain Mineral Law Institute, 24th, 1978 Proceedings. New York: Matthew Bender & Company, 1978, 811-831.

Errebo, Burns H. "Coal Royalties." in Rocky Mountain Mineral Law Institute, 26th, 1980 Proceedings. New York: Matthew Bender & Company, 1980, 75-115.

Fletcher, F. Alan. "Mining Leases and Oil and Gas Leases--Different Breeds," in Rocky Mountain Mineral Law Institute, 24th, 1978 Proceedings. New York: Matthew Bender & Company, 1978, 309-346.

Klein, Blair, and Burke, Noel. "Farmout Agreement: Its Form and Substance," in Rocky Mountain Mineral Law Institute, 24th, 1978 Proceedings. New York: Matthew Bender & Company, 1978, 479-520.

Mining Journal Books Ltd. Negotiation & Drafting of Mining Development Agreements. England: Mining Journal, 1980.

Parcel, Randy L. "Selected Royalty Problems of Standard Mining Lease Forms," in Rocky Mountain Mineral Law Institute, 19th, 1973 Proceedings. New York: Matthew Bender & Company, 1974, 367-392.

Peterson, Royal E. "The Uranium Royalty Provision: Its Evolution, Present Complexity and Future Uncertainty," in Rocky Mountain Mineral Law Institute, 22nd, 1976 Proceedings. New York: Matthew Bender & Company, 1976, 865-889.

Scott, Jeffrey J. "Coal Supply Agreements," in Rocky Mountain Mineral Law Institute, 23rd, 1977 Proceedings. New York: Matthew Bender & Company, 1977, 107-172.

Sebenius, James K., and Pal, Mati L. "Evolving Financial Terms of Mineral Agreements: Risks, Rewards and Participation in Deep Seabed Mining." Columbia Journal of World Business 15 (Winter 1980): 75-83.

Stroud, Garner R. "Acquisition of Coal Rights--It's Different," in *Rocky Mountain Mineral Law Institute*, 21st, 1975 Proceedings. New York: Matthew Bender & Company, 1975, 587-615.

News Releases and Short Comments

"Mining and Land Deals." *Accountant* (Eng.) 170 (May 9, 1974): 587.

JOINT VENTURES

Grandey, Gerald W. "Joint Venture Agreements Between Coal Companies and Other Participant Variations," in *Rocky Mountain Mineral Law Institute*, 25th, 1979 Proceedings. New York: Matthew Bender & Company, 1979, 6-1--6-52.

McDonald, Alan A. "Joint Venture Accounting," in *Rocky Mountain Mineral Law Institute*, 20th, 1974 Proceedings. New York: Matthew Bender & Company, 1975, 541-563.

Peterson, Royal E. "Decision-Making in Joint Ventures," in *Rocky Mountain Mineral Law Institute*, 24th, 1978 Proceedings. New York: Matthew Bender & Company, 1978, 453-478.

MERGERS AND ACQUISITIONS

Allen, Zachariah. "Who Is Captive When Utilities Buy Coal Companies?" *Public Utilities Fortnightly* 103 (January 4, 1979): 16-20.

Brooks, LeRoy D., and Rhyne, R. Glenn. "Utility Performance with Captive Coal Companies." *Public Utilities Fortnightly* 108 (July 2, 1981): 19-24.

Continental Oil Company. *Oil, Coal, and Uranium: Do They Mix?* Stamford, Conn., 1977.

Gransee, Marsha F. "Regulation of Captive Coal Costs." *Public Utilities Fortnightly* 107 (January 29, 1981): 46-50.

Massella, Ines M. "Repricing Affiliated Coal--An Update." *Public Utilities Fortnightly* 110 (December 9, 1982): 58-61.

Wang, Paul. "Leveraged Buy-Out Financing of a Coal Mine." Mergers and Acquisitions 12 (Fall 1977): 34-45.

News Releases and Short Comments

"Captive Customers? Utility-Owned Mines, Meant to Assure Fuel, Often Lift Power Cost; Critics Say Some Firms Run Mines Badly or Use Them to Escape Profit Limits; Regulators Begin to Notice." Wall Street Journal, 10 May 1979, 1.

"Mining Companies Become Targets of Oil Industry's Drive to Diversity." Wall Street Journal, 13 March 1981, 25.

"Mines Over Matter: Editorial Concerning Big Oil Companies Diversifying Into Mining." Wall Street Journal, 17 March 1981, 32.

"Letter to the Editor Concerning March 17 Editorial Justifying the Purchase of Mining Companies by Oil Companies." Wall Street Journal, 9 April 1981, 29.

"Are Takeovers Good for Mining." Wall Street Journal, 13 May 1981, 26.

"Lure of Coal Reserves Diminishes for Some Acquisitive Companies." Wall Street Journal, 22 February 1982, 25.

ECONOMICS, FINANCE, AND EVALUATION

Anders, Gerhard, et al., eds. The Economics of Mineral Extraction. New York: Praeger Publishers, 1980.

Bloomenthal, Harold S. "Financing Mineral Ventures," in Rocky Mountain Mineral Law Institute, 23rd, 1977 Proceedings. New York: Matthew Bender & Company, 1977, 257-352.

Carlson, Rodger D. The Economics for the Profitable Mining & Marketing of Gold, Silver, Copper, Lead & Zinc Ores. Lanham, Md.: University Press of America, 1982.

Clark, Graham M. "Economic Considerations for the Layman/Landman in Negotiating Mining Leases," in Rocky Mountain Mineral Law Institute, 24th, 1978 Proceedings. New York: Matthew Bender & Company, 1978, 811-831.

David, M. Geostatistical Ore Reserve Estimation. Developments in Geomathematics, Vol. 2. New York: Elsevier Science Publishing Co., 1977.

Drechsler, Herbert D., and Wane, Malcolm T. "Mining Industry and Scarcity of Minerals." Columbia Journal of World Business 9 (Fall 1974): 65-70.

Gentry, Donald W. "Valuation of Mines and Mineral-Bearing Lands." Valuation 27 (November 1981): 28-38.

Goldie, B. W. "Merchant Banking: Project Finance--Natural Resource Style." Chartered Accountant in Australia' 51 (December 1980): 27, 30-31.

Haas, Albert. "Switching to Coal Is Buying Trouble." Business and Society Review (Winter 1974-75): 52-57.

Heal, Geoffrey. "Relationship Between Price and Extraction Cost for a Resource with a Backstop Technology." Bell Journal of Economics 7 (Autumn 1976): 371-378.

King, Joseph E. A Mine to Make A Mine: Financing the Colorado Mining Industry. College Station: Texas A & M University Press, 1977.

Lee, Roy S. "Enterprise: Operational Aspects and Implications," Columbia Journal of World Business 15 (Winter 1980): 62-74.

Neild, Paul. "Can We Really Afford the Miners." Accountancy (Eng.) 85 (July 1974): 50-51, 53-54.

Paschall, Robert H. "Appraisal of Mineral-Producing Properties." Valuation 21 (October 1974): 2-9.

Phillips, Peter J., and De Rienzo, Paul P. "Fixing Up Coal: What It Can Cost." Public Utilities Fortnightly 99 (April 14, 1977): 29-33.

Radetzki, Marian, and Zorn, Stephen. Financing Mining Projects in Developing Countries: A United Nations Study. New York: International Pubns. Service, 1979.

Rendu, Jo M. Introduction to Geostatistical Methods of Mineral Evaluation. England: Mining Journal, 1980.

Sassoon, David M. "Financing the Mining Sector: The Bank's Role." Finance & Development 12 (September 1975): 21-23, 41.

Schlottmann, Alan, and Spore, Robert L. "Economic Impacts of Surface Mine Reclamation." Land Economics 52 (August 1976): 265-277.

Shekhar, A. K. Chandra. "Economic Working of Mines." Management Accountant (India) 9 (February 1974): 129-132.

Sibley, Angus. "Coal: Twenty Years of Investment Potential." Accountant's Magazine (Scot.) 85 (December 1981): 424-426.

Sirola, D. Grant. "Canadian Mining Industry in Prospective." CA Magazine (Canada) 102 (January 1973): 63-64.

Summers, Kenneth R. "Financing Expanded Coal Production." Banking 69 (July 1977): 84.

Summers, Kenneth R. "Lending to the Coal Industry." Journal of Commercial Bank Lending 61 (November 1978): 8-18.

Swieringa, Robert J. "The Silver-Lined Bonds of Sunshine Mining." Accounting Review 56 (January 1981): 166-176.

Taylor, John D. "Valuation of a Mining Company--An Overview." South African Chartered Accountant 12 (August 1976): 271-275.

Wang, Paul. "Leveraged Buy-Out Financing of a Coal Mine." Mergers and Acquisitions 12 (Fall 1977): 34-45.

Whitman, George E. "State Industry at the Crossroads--III." Certified Accountant (Eng.) (February 1975): 67-68, 74-76.

Wynant, Larry. "Project Financing for Extractive Ventures." Management Accounting 60 (October 1978): 29-35.

News Releases and Short Comments

Strain, William J., and Braley, John A. "Stormy Weather for the Mining Industry." CA Magazine (Canada) 106 (January 1975): 37-42.

U.S. Department of Commerce. "Industries Keep Sharper Eye on Non-Ferrous Metal Supply." Commerce Today 5 (April 28, 1975): 18-19.

"Florida Will File Suit Against Four Coal Mining Companies and Six Individuals, Alleging They Defrauded Investors Through Sale of Limited Partnerships in Coal-Mining Operations." Wall Street Journal, 14 July 1977, 13.

"SEC Warned Prospective Investors in Coal-Mining Ventures That They May Be Defrauded if They Don't Take a Close Look at the Proposals," Wall Street Journal, 16 December 1977, 10.

"SEC Staff Charged Lebkeucher, Spencer, Inc., Sarasota, Fla., With Violating Securities Laws in the Sale of Limited Partnerships Formed to Mine and Sell Coal. COAL & PARTNERSHIPS & MINING." Wall Street Journal, 12 December 1982, 2.

STATISTICS

David, M. Geostatistical Ore Reserve Estimation. Developments in Geomathematics, Vol. 2. New York: Elsevier Science Publishing Co., 1977.

Guarascio, Massimo, et al., eds. Advanced Geostatistics in the Mining Industry. Nato Mathematical & Physical Sciences Series, No. C24. Holland: Reidel, 1976.

Journel, A. G., and Huijbregts, C. J. Mining Geostatistics. New York: Academic Press, 1979.

Rendu, Jo M. Introduction to Geostatistical Methods of Mineral Evaluation. England: Mining Journal, 1980.

REGULATION

Achterman, Gail L. "Cost Recovery by Federal Land Management Agencies: Fees for Benefits Received or Unlawful Tax?" in Rocky Mountain Mineral Law Institute, 26th, 1980 Proceedings. New York: Matthew Bender & Company, 1980, 313-364.

Bulow, Richard P. "Government Regulation of the Coal Industry." Pennsylvania CPA Spokesman 49 (April 1979): 6-8.

Gazda, Steven R., and Maccaroni, Ronald J. "Auditing the Coal Mine Dust Sampling Program." GAO Review 10 (Fall 1976): 47-53.

Gransee, Marsha F. "Regulation of Captive Coal Costs." Public Utilities Fortnightly 107 (January 29, 1981): 46-50.

Parcel, Randy L. "Federal, State and Local Regulation of Mining Exploration," in Rocky Mountain Mineral Law Institute, 22nd, 1976 Proceedings. New York: Matthew Bender & Company, 1976, 405-432.

Rothenberg, Jerome. "Impact of Regulation on an Exhaustive Resource Industry: A Methodological Approach and A Model" Law and Contemporary Problems 43 (Winter/Spring 1979): 112-149.

Schlauch, Paul J. "Tripartite Federalism--The Emerging Role of Local Government as a Regulator of the Extractive Industries," in Rocky Mountain Mineral Law Institute, 20th, 1974 Proceedings. New York: Matthew Bender & Company, 1975, 359-397.

News Releases and Short Comments

"Florida Will File Suit Against Four Coal Mining Companies and Six Individuals, Alleging They Defrauded Investors Through Sale of Limited Partnerships in Coal-Mining Operations." Wall Street Journal, 14 July 1977, 13.

"SEC Warned Prospective Investors in Coal-Mining Ventures That They May Be Defrauded if They Don't Take a Close Look at the Proposals," Wall Street Journal, 16 December 1977, 10.

Walker, Michael A. "Form S-18 Amended for Mining Companies and Form S-3 Rescinded." CPA Journal 51 (September 1981): 83.

"SEC Staff Charged Lebkeucher, Spencer, Inc., Sarasota, Fla., With Violating Securities Laws in the Sale of Limited Partnerships Formed to Mine and Sell Coal. COAL & PARTNERSHIPS & MINING." Wall Street Journal, 12 December 1982, 2.

BLACK LUNG

Amoroso, Vincent. "Black Lung: A Dark Liability." Pension World 14 (December 1978): 40-42.

Coe, Teddy L. "Accounting for Black Lung Benefit Costs." Journal of Extractive Industries Accounting 1 (Summer 1982): 59-66.

Viers, Jimmy D. "Trust Fund for Payment of Black Lung Benefits." Virginia Accountant 32 (September 1980): 19-21.

News Releases and Short Comments

"Black Lung Victims Still Get Paid by the Government, Despite the Law; a Year After the Legal Deadline for Mine Operators to Take Over Benefit Payments for New Black Lung Claimants, the Federal Treasury Is Still Doling Out the Funds." Wall Street Journal, 14 January 1975, 1.

"Carter Opposed Black-Lung Benefit Plan of Union but Would Ease Rules on Proof, Labor Dept. Spokesman Said." Wall Street Journal, 18 March 1977, 7.

"House Labor Committee Voted for Major Revision of National Black Lung Benefit Program for Coal Miners and Their Families." Wall Street Journal, 23 March 1977, 32.

"Senate Cleared Measure Which Will Make It Easier for Coal Miners to Qualify for Black Lung Benefits." Wall Street Journal, 21 September 1977, 19.

"The House Gave Final Congressional Approval to a Bill That Makes It Easier for Coal Miners With Black Lung Disease to Get Financial Compensation and Allows Them to Reapply for Benefits Denied in the Past." Wall Street Journal, 16 February 1978, 5.

"Feb. 28, Leaders of the United Mine Workers Union Approved a Previously Proposed National Coal Walkout to Protest President Reagan's Plan to Cut Back Federal Subsidies to the Black-Lung Trust Fund." Wall Street Journal, 2 March 1981, 12.

"Black Lung Prognosis: Editorial Concerning the Reagan Administration's Goal to Restore the Black Lung Program to What Congress Originally Intended: A Set of Benefits for Miners with Damaged Health." Wall Street Journal, 18 March 1981, 28.

"Black Lung Chances Hit A Snag; the Dragged-Out Coal Miners' Strike." Wall Street Journal, 28 April 1981, 1.

PRICING

Carney, Edmund M. "Pricing Provisions in Coal Contracts," in Rocky Mountain Mineral Law Institute, 24th, 1978 Proceedings. New York: Matthew Bender & Company, 1978, 197-232.

Heal, Geoffrey. "Relationship Between Price and Extraction Cost for a Resource with a Backstop Technology." Bell Journal of Economics 7 (Autumn 1976): 371-378.

Massella, Ines M. "Repricing Affiliated Coal--An Update." Public Utilities Fortnightly 110 (December 9, 1982): 58-61.

McNicol, David L. "Two Price Systems in the Copper Industry." Bell Journal of Economics 6 (Spring 1975): 50-73.

PLANNING, CONTROL, AND BUDGETING

Alonzo, Martin V. "Corporate Strategy for Combating Inflation." Management Accounting 59 (March 1978): 57-60.

Ang, Po Tim T. "Developing a Computer Based Responsibility Accounting System for a Mining Operation." SGV Group Journal 1 (1980): 68-73.

Bhattacharyya, P. S. "Production and Productivity in Coal Industry." Management Accountant (India) 15 (June 1980): 243- 244, 247.

Cline, Robert L. Directory of Computer Models Applicable to the Coal Mining Industry. Golden, Colo.: Colorado School of Mines, 1981.

Deb, N. G. "Productivity in Coal Industry and General Consideration." Management Accountant (India)" (April 1, 1976): 286-294.

Ezra, Derek. "Long Term Planning for Coal." Long Range Planning (December 1974): 21-23.

Lewy, P. M. "Colliery Model: A Computer Aid to Planning at a Colliery." South African Chartered Accountant 16 (July 1980): 298-304.

Nath, K. "Cost Control and Cost Reduction in Coal Industry." Management Accountant (India) 14 (October 1979): 919-926.

Navin, Thomas R. Copper Mining & Management. Tucson: University of Arizona Press, 1978.

Patterson, A. L. "Taking Inventory with a Camera." GAO Review (Fall 1974): 32-36.

Potgieter, Theunis Johannes Eugune: "The Measurement of Divisional Performance in the Gold-Mining Industry." D. Com. dissertation, University of South Africa, 1979.

Prasad, B. M. "Fallacy of Productivity in Coal Industry--A Case Study." Management Accountant (India) 13 (April 1978): 379.

Rao, G. Gopala. "Costing and Cost Control in Iron Ore Mining Industry." Management Accountant (India) 16 (December 1981): 577-579.

Sedgwick, Adam. "Keeping It Simple: A Case Study in Corporate Planning." Long Range Planning (Eng.) 11 (February 1978): 10-14.

Sekhar, R. C. "Manpower Wastage and Control in Coal Industry: New Thoughts and Statistical Methods." Management Accountant (India) 14 (October 1979): 936-941.

Symposium On Coal Management Techniques, 2nd, National Coal Association and Bituminous Coal Research, 1976 Papers. Washington: National Coal Association, 1977. (Includes "Results of Survey on Accounting Principles and Procedures Practiced in the Coal Industry.")

Symposium On Coal Management Techniques, 3rd, National Coal Association and Bituminous Coal Research, 1977 Papers. Washington: National Coal Association, 1978.

Wright, V. C. "Corporate Planning in the Mineral Industry." Long Range Planning 9 (April 1976): 19-23.

News Releases and Short Comments

"Guaranteed Cost Insurance Has Drawbacks." Business Insurance 16 (October 11, 1982): 67.